Synthesis Lectures on Engineering, Science, and Technology

The focus of this series is general topics, and applications about, and for, engineers and scientists on a wide array of applications, methods and advances. Most titles cover subjects such as professional development, education, and study skills, as well as basic introductory undergraduate material and other topics appropriate for a broader and less technical audience.

C. Y. Wang

Essential Analytic Laminar Flow

 Springer

C. Y. Wang
Michigan State University
East Lansing, MI, USA

ISSN 2690-0300 ISSN 2690-0327 (electronic)
Synthesis Lectures on Engineering, Science, and Technology
ISBN 978-3-031-36448-8 ISBN 978-3-031-36449-5 (eBook)
https://doi.org/10.1007/978-3-031-36449-5

This Springer imprint is published by the registered company Springer Nature Switzerland AG
The registered company address is: Gewerbestrasse 11, 6330 Cham, Switzerland

Preface

Fluid mechanics encompasses such a wide variety of subjects that nowadays it is almost impossible for anyone to write a book and do justice to all the topics. Tomes of hundreds of pages have been written on just one major sub-topic, such as compressible flow, turbulence, computational fluid dynamics, and experimental methods. Even then, it is difficult to proclaim the presentation as complete.

Thus, it is essential to state the aim, the limitations, and the philosophy of the present work at the outset.

- Aim: To present essential fluid mechanics to students and researchers, just enough to do independent analytical work.
- Limitations: Only include the analytical (not numerical, experimental, or empirical) methods and solutions of the constant property Navier-Stokes equation and their closely related applications.
- Philosophy: To achieve the goal as directly as possible, leaving some non-essential details to the references.

There are different levels in presenting fluid mechanics. The introductory level considers statics, control volume, etc. The advanced level uses tensors, theorems about existence, etc. The present book is in the intermediate level for the early career researcher or the first-year graduate student. For best results, the reader should have had undergraduate differential equations, some fluid mechanics exposure, and know how to use simple computer software and the Science Citation Index (forwards and backwards) for additional references.

Suggested flowchart for this book is as follows: Chap. 1 "The Navier-Stokes Equation", Chap. 2 "Exact Solutions" (perhaps supplemented by Appendix A on similarity methods), Chap. 3 "Non-dimensionalization, Scaling and Approximations", Chap. 4 "Boundary Layers" (perhaps supplemented by Appendix B Perturbation Theory and Appendix C Potential Flow). Then the reader is free to choose from among the special topics in Chaps. 5–9.

In comparison to other viscous flow texts, the present work differs in the following respects:

- This book is short and concise.
- Topics and exercises are presented to the verge of original research.
- It is simple enough for self-study.
- The method is illustrated in the examples.
- The use of stream function is emphasized.

I have done teaching and research in various aspects of fluid mechanics for the past 50+ years, and this book would no doubt reflect some of my personal preferences.

Reporting any mistakes or inadequacies would be appreciated.

East Lansing, USA   C. Y. Wang
2023

Contents

About the Author

C. Y. Wang is a professor in the Department of Mathematics and an adjoint professor in the Department of Mechanical Engineering at Michigan State University, East Lansing, Michigan. He obtained his B.S. from Taiwan University and Ph.D. from Massachusetts Institute of Technology. Professor Wang published about 330 papers in fluid mechanics and thermos-fluids. He wrote a book *Essential Perturbation Methods* (Springer-Nature) and is a coauthor of *Exact Solutions for Buckling of Structural Members* (CRC Press), and *Structural Vibrations* (CRC Press). He has served as a technical editor of *Applied Mechanics Reviews*.

The Navier–Stokes Equation

1

The governing equation of viscous flow is the Navier–Stokes (N–S) equation. In this chapter we shall derive the N–S equation with the least effort possible.

The assumptions are that the fluid is a continuum (can be differentiated), isotropic (no preferred direction), Newtonian (stress proportional to strain rate) and its properties (density, viscosity) are constant.

We shall derive the N–S equation in two-dimensional Cartesian coordinates, extend to three dimensions, and finally to general orthogonal curvilinear coordinates. The choice of a proper coordinate system is important since it is necessary for analytically solving specific problems.

1.1   Deriving the N–S Equation

Consider first the continuity equation. Let (u, v) be two-dimensional velocities in the Cartesian (x, y) directions respectively.

Figure 1.1a shows an elemental area $\Delta x \Delta y$ where $u(x, y)$ enters and when it leaves on the other side becomes $u(x + \Delta x, y)$, similarly for the v velocity component. Since the density is constant, the net fluid loss is zero.

$$[u(x + \Delta x, y) - u(x, y)]\Delta y + [v(x, y + \Delta y) - v(x, y)]\Delta x = 0 \tag{1.1}$$

Expanding in Taylor series

$$u(x + \Delta x, y) = u(x, y) + \Delta x \frac{\partial u}{\partial x}(x, y) + \cdots \tag{1.2}$$

Equation (1.1) simplifies to

© The Author(s), under exclusive license to Springer Nature Switzerland AG 2024
C. Y. Wang, *Essential Analytic Laminar Flow*, Synthesis Lectures on Engineering, Science, and Technology, https://doi.org/10.1007/978-3-031-36449-5_1

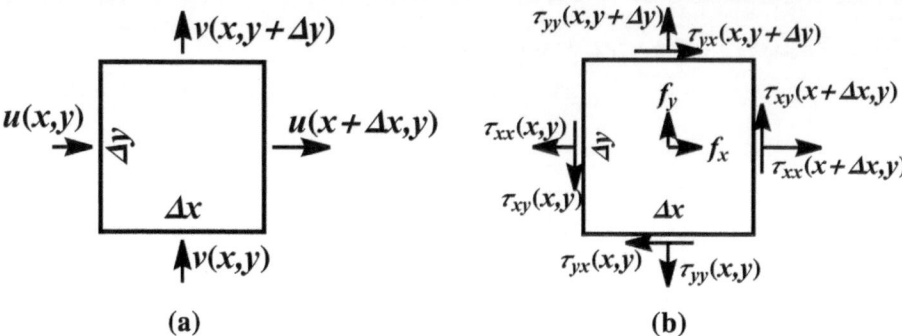

Fig. 1.1 An elemental area **a** velocities through the boundary **b** stresses and forces

$$\frac{\partial u}{\partial x} + \frac{\partial v}{\partial y} = 0 \tag{1.3}$$

Extending to three dimensions and in vector form the continuity equation is

$$\nabla \cdot \boldsymbol{u} = 0 \tag{1.4}$$

where $\boldsymbol{u}$ is the velocity vector. Equation (1.4) is also valid when the flow is unsteady (time dependent).

Let $\boldsymbol{u}(\boldsymbol{x}, t)$ be the velocity vector, t be the time and $\boldsymbol{x}$ be the spatial (Eulerian) position. $\boldsymbol{x}$ is function of some original position $\boldsymbol{\xi}$ and t. Thus, the acceleration following the mass (Lagrangian) is

$$\boldsymbol{a} = \frac{d}{dt}\boldsymbol{u}\big[\boldsymbol{x}(\boldsymbol{\xi}, t), t\big] = \frac{\partial \boldsymbol{u}}{\partial t} + \sum \frac{\partial \boldsymbol{u}}{\partial x_i}\frac{dx_i}{dt} = \frac{\partial \boldsymbol{u}}{\partial t} + (\boldsymbol{u} \cdot \nabla)\boldsymbol{u} \tag{1.5}$$

The forces on an elemental volume of fluid consist of body forces which act on the whole mass and surface forces which act on the bonding surface. Figure 1.1b shows a two-dimensional elemental area subjected to body forces per area (f_x, f_y) and stresses $(\tau_{xx}, \tau_{xy}, \tau_{yx}, \tau_{yy})$. Similar to Eq. (1.1) the net force in the x direction is

$$F_x = \frac{\partial \tau_{xx}}{\partial x}\Delta x \Delta y + \frac{\partial \tau_{yx}}{\partial y}\Delta y \Delta x + f_x \Delta x \Delta y \tag{1.6}$$

Since Newton's momentum law holds, the mass times acceleration is equal to the net force. Using the x component of Newton's law gives

$$\rho \Delta x \Delta y \left[\frac{\partial u}{\partial t} + (\boldsymbol{u} \cdot \nabla)u\right] = F_x \tag{1.7}$$

or

$$\rho\left[\frac{\partial u}{\partial t} + (\boldsymbol{u} \cdot \nabla)u\right] = \frac{\partial \tau_{xx}}{\partial x} + \frac{\partial \tau_{yx}}{\partial y} + f_x \tag{1.8}$$

Similarly in the y direction

$$\rho\left[\frac{\partial v}{\partial t} + (\boldsymbol{u} \cdot \nabla)v\right] = \frac{\partial \tau_{yy}}{\partial y} + \frac{\partial \tau_{xy}}{\partial x} + f_y \tag{1.9}$$

A moment balance of the elemental area shows the stress (tensor) is symmetric.

$$\tau_{xy} = \tau_{yx} \tag{1.10}$$

Define a strain rate tensor

$$d_{ij} = \frac{1}{2}\left(\frac{\partial u_i}{\partial x_j} + \frac{\partial u_j}{\partial x_i}\right) \tag{1.11}$$

For an isotropic Newtonian fluid (most fluids) the stress is proportional to strain rate

$$\tau_{ij} = -p\delta_{ij} + 2\mu d_{ij} \tag{1.12}$$

Here δ_{ij} is the Kronecker delta, p is the pressure and μ is the viscosity. In two dimensions

$$\tau_{xx} = -p + 2\mu\frac{\partial u}{\partial x}, \quad \tau_{xy} = \mu\left(\frac{\partial u}{\partial y} + \frac{\partial v}{\partial x}\right), \quad \tau_{yy} = -p + 2\mu\frac{\partial v}{\partial y} \tag{1.13}$$

Upon substituting into Eq. (1.8) and using Eq. (1.3), we obtain

$$\rho\left[\frac{\partial u}{\partial t} + (\boldsymbol{u} \cdot \nabla)u\right] = -\frac{\partial p}{\partial x} + \mu\left(\frac{\partial^2 u}{\partial x^2} + \frac{\partial^2 u}{\partial y^2}\right) + f_x \tag{1.14}$$

Extending to three dimensions yields the N–S equation

$$\frac{\partial \boldsymbol{u}}{\partial t} + (\boldsymbol{u} \cdot \nabla)\boldsymbol{u} = -\frac{1}{\rho}\nabla p + \nu\nabla^2\boldsymbol{u} + \boldsymbol{f} \tag{1.15}$$

where $\boldsymbol{f}$ is the body force per mass and $\nu = \mu/\rho$ is the kinematic viscosity. Equation (1.15) is supplemented by continuity Eq. (1.4), initial condition and boundary conditions.

The three-dimensional form of the N–S equation is seldom used in analytic work. Here we present the two-dimensional form in Cartesian coordinates.

$$\frac{\partial u}{\partial t} + u\frac{\partial u}{\partial x} + v\frac{\partial u}{\partial y} = -\frac{1}{\rho}\frac{\partial p}{\partial x} + \nu\left(\frac{\partial^2 u}{\partial x^2} + \frac{\partial^2 u}{\partial y^2}\right) + f_x \tag{1.16}$$

$$\frac{\partial v}{\partial t} + u\frac{\partial v}{\partial x} + v\frac{\partial v}{\partial y} = -\frac{1}{\rho}\frac{\partial p}{\partial y} + \nu\left(\frac{\partial^2 v}{\partial x^2} + \frac{\partial^2 v}{\partial y^2}\right) + f_y \tag{1.17}$$

From Eq. (1.3) a stream function ψ can be defined

$$u = \frac{\partial \psi}{\partial y}, \quad v = -\frac{\partial \psi}{\partial x} \tag{1.18}$$

Then eliminating pressure, the N–S equation becomes

$$\frac{\partial}{\partial t}\nabla^2\psi + \frac{\partial \psi}{\partial y}\frac{\partial}{\partial x}\nabla^2\psi - \frac{\partial \psi}{\partial x}\frac{\partial}{\partial y}\nabla^2\psi = \nu\nabla^4\psi + \left(\frac{\partial f_x}{\partial y} - \frac{\partial f_y}{\partial x}\right) \tag{1.19}$$

In two dimensions, the vorticity is

$$\zeta = \nabla^2\psi \tag{1.20}$$

Notice, if the body forces f_x, f_y are conservative (such as gravity), then they can be absorbed into the pressure term and the last parenthesis in Eq. (1.19) can be set to zero. In terms of a Jacobian, the N–S equation without body forces is

$$\frac{\partial}{\partial t}\nabla^2\psi + \frac{\partial(\nabla^2\psi, \psi)}{\partial(x, y)} = \nu\nabla^4\psi \tag{1.21}$$

Equation (1.21) is a partial differential equation to be solved with an initial condition, four boundary conditions in x, and four boundary conditions in y. On solid boundaries we assume the no-slip condition, unless partial slip occurs (see Chap. 7).

1.2   N–S Equation in Other Coordinates

A coordinate system that fits (or closely fits), the boundary is necessary for obtaining an analytic solution.

For non-orthogonal coordinate systems, tensors and Christoffel symbols are necessary. An example is the flow in a helical tube with non-zero pitch.

For orthogonal curvilinear coordinate systems, let $x(\xi)$ denote the relationship between the new system ξ and the Cartesian system x. Construct the elemental distance squared

$$ds^2 = dx \cdot dx = \sum_{i=1}^{3} h_i^2 d\xi_i^2 \tag{1.22}$$

where h_i are scale factors, and let a_i be the unit vectors in the directions ξ_i of the curvilinear system. Equation (1.22) also ascertains the curvilinear system is orthogonal. Vector calculus shows

$$\nabla\phi = \sum_{1}^{3} \frac{a_i}{h_i}\frac{\partial \phi}{\partial \xi_i} \tag{1.23}$$

$$\nabla^2 \phi = \frac{1}{h_1 h_2 h_3} \sum_1^3 \frac{\partial}{\partial \xi_i} \left(\frac{h_{i+1} h_{i+2}}{h_i} \frac{\partial \phi}{\partial \xi_i} \right) \tag{1.24}$$

where the index i is mod[3], i.e. $h_4 = h_1, h_5 = h_2$. Any vector can be decomposed

$$\boldsymbol{F} = \sum_1^3 F_i \boldsymbol{a}_i \tag{1.25}$$

then

$$\nabla \cdot \boldsymbol{F} = \frac{1}{h_1 h_2 h_3} \sum_1^3 \frac{\partial}{\partial \xi_i} (h_{i+1} h_{i+2} F_i) \tag{1.26}$$

$$\nabla \times \boldsymbol{F} = \frac{1}{h_1 h_2 h_3} \begin{vmatrix} h_1 \boldsymbol{a}_1 & h_2 \boldsymbol{a}_2 & h_3 \boldsymbol{a}_3 \\ \frac{\partial}{\partial \xi_1} & \frac{\partial}{\partial \xi_2} & \frac{\partial}{\partial \xi_3} \\ h_1 F_1 & h_2 F_2 & h_3 F_3 \end{vmatrix} \tag{1.27}$$

$$(\boldsymbol{G} \cdot \nabla)\boldsymbol{F} = \sum_1^3 \boldsymbol{a}_i \left[\begin{array}{l} (\boldsymbol{G} \cdot \nabla) F_i + \dfrac{F_{i+1}}{h_i h_{i+1}} \left(G_i \dfrac{\partial h_i}{\partial \xi_{i+1}} - G_{i+1} \dfrac{\partial h_{i+1}}{\partial \xi_i} \right) \\[3mm] + \dfrac{F_{i+2}}{h_i h_{i+2}} \left(G_i \dfrac{\partial h_i}{\partial \xi_{i+2}} - G_{i+2} \dfrac{\partial h_{i+2}}{\partial \xi_i} \right) \end{array} \right] \tag{1.28}$$

Notice for Eq. (1.15) the cross product should be used

$$\nabla^2 \boldsymbol{u} = \nabla(\nabla \cdot \boldsymbol{u}) - \nabla \times (\nabla \times \boldsymbol{u}) = -\nabla \times (\nabla \times \boldsymbol{u}) \tag{1.29}$$

where Eq. (1.4) has been applied. Typical strain rate tensors are

$$d_{11} = \frac{1}{h_1} \frac{\partial u_1}{\partial \xi_1} + \frac{u_2}{h_1 h_2} \frac{\partial h_1}{\partial \xi_2} + \frac{u_3}{h_1 h_3} \frac{\partial h_1}{\partial \xi_3} \tag{1.30}$$

$$d_{12} = \frac{h_2}{2h_1} \frac{\partial}{\partial \xi_1} \left(\frac{u_2}{h_2} \right) + \frac{h_1}{2h_2} \frac{\partial}{\partial \xi_2} \left(\frac{u_1}{h_1} \right) \tag{1.31}$$

1.2.1 Cylindrical Coordinates

Let (u, v, w) be velocities in the directions of the cylindrical coordinates (r, θ, z). Since $x = r \cos(\theta)$, $y = r \sin(\theta)$, $z = z$, Eq. (1.22) gives $h_1 = 1, h_2 = r, h_3 = 1$. For two dimensional flow independent of the z direction, the N–S equation (without the body force terms) is

$$\frac{\partial u}{\partial t} + u \frac{\partial u}{\partial r} + \frac{v}{r} \frac{\partial u}{\partial \theta} - \frac{v^2}{r} = -\frac{1}{\rho} \frac{\partial p}{\partial r} + \nu \left(\nabla^2 u - \frac{u}{r^2} - \frac{2}{r^2} \frac{\partial v}{\partial \theta} \right) \tag{1.32}$$

$$\frac{\partial v}{\partial t} + u\frac{\partial v}{\partial r} + \frac{v}{r}\frac{\partial v}{\partial \theta} + \frac{uv}{r} = -\frac{1}{\rho r}\frac{\partial p}{\partial \theta} + \nu\left(\nabla^2 v - \frac{v}{r^2} + \frac{2}{r^2}\frac{\partial u}{\partial \theta}\right) \tag{1.33}$$

The relevant strain rate useful for shear stress is

$$d_{r\theta} = \frac{1}{2}\left(\frac{\partial v}{\partial r} - \frac{v}{r} + \frac{1}{r}\frac{\partial u}{\partial \theta}\right) \tag{1.34}$$

The continuity equation is satisfied by

$$u = \frac{1}{r}\frac{\partial \psi}{\partial \theta}, \quad v = -\frac{\partial \psi}{\partial r} \tag{1.35}$$

The N–S equation reduces to

$$\frac{\partial}{\partial t}\nabla^2\psi + \frac{1}{r}\frac{\partial(\nabla^2\psi, \psi)}{\partial(r, \theta)} = \nu\nabla^4\psi \tag{1.36}$$

where

$$\nabla^2 = \frac{\partial^2}{\partial r^2} + \frac{1}{r}\frac{\partial}{\partial r} + \frac{1}{r^2}\frac{\partial^2}{\partial \theta^2} \tag{1.37}$$

On the other hand, if the flow is axisymmetric about the z axis with a possible swirl, and independent of the angle θ, let (u, v, w) be velocities in the (r, θ, z) directions respectively. The N–S equation is

$$\frac{\partial u}{\partial t} + u\frac{\partial u}{\partial r} + w\frac{\partial u}{\partial z} - \frac{v^2}{r} = -\frac{1}{\rho}\frac{\partial p}{\partial r} + \nu\left(\nabla^2 u - \frac{u}{r^2}\right) \tag{1.38}$$

$$\frac{\partial w}{\partial t} + u\frac{\partial w}{\partial r} + w\frac{\partial w}{\partial z} = -\frac{1}{\rho}\frac{\partial p}{\partial z} + \nu\left(\nabla^2 w\right) \tag{1.39}$$

$$\frac{\partial v}{\partial t} + u\frac{\partial v}{\partial r} + w\frac{\partial v}{\partial z} + \frac{uv}{r} = -\frac{1}{\rho r}\frac{\partial p}{\partial \theta} + \nu\left(\nabla^2 v - \frac{v}{r^2}\right) \tag{1.40}$$

$$\nabla^2 = \frac{\partial^2}{\partial r^2} + \frac{1}{r}\frac{\partial}{\partial r} + \frac{\partial^2}{\partial z^2} \tag{1.41}$$

The strain rate is

$$d_{rz} = \frac{1}{2}\left(\frac{\partial u}{\partial z} + \frac{\partial w}{\partial r}\right) \tag{1.42}$$

From continuity define a stream function

$$w = \frac{1}{r}\frac{\partial \psi}{\partial r}, \quad u = -\frac{1}{r}\frac{\partial \psi}{\partial z} \tag{1.43}$$

The N–S equation becomes

$$\frac{\partial}{\partial t}E^2\psi - \frac{1}{r}\frac{\partial(E^2\psi,\psi)}{\partial(r,z)} + \frac{2}{r^2}E^2\psi\frac{\partial\psi}{\partial z} + 2v\frac{\partial v}{\partial z} = vE^4\psi \tag{1.44}$$

$$\frac{\partial}{\partial t}(rv) + \frac{1}{r}\frac{\partial(\psi,rv)}{\partial(r,z)} = vE^2(rv) \tag{1.45}$$

where

$$E^2 = \frac{\partial^2}{\partial r^2} - \frac{1}{r}\frac{\partial}{\partial r} + \frac{\partial^2}{\partial z^2} \tag{1.46}$$

1.2.2   Spherical Coordinates

For spherical coordinates, let (u, v, w) be the velocities in the spherical directions $(\varrho, \theta, \varphi)$, where ϱ is the radial distance from the origin, θ is the angle between the ϱ vector to the z axis, and φ is the angle of the ϱ, z plane. Let the flow be axisymmetric about the z axis. We find $h_1 = 1, h_2 = \varrho, h_3 = \varrho\sin(\theta)$. The N–S equation becomes

$$\frac{\partial u}{\partial t} + u\frac{\partial u}{\partial\varrho} + \frac{v}{\varrho}\frac{\partial u}{\partial\theta} - \frac{(v^2+w^2)}{\varrho} = -\frac{1}{\rho}\frac{\partial p}{\partial\varrho} + v\left(\nabla^2 u - \frac{2u}{\varrho^2} - \frac{2}{\varrho^2}\frac{\partial v}{\partial\theta} - \frac{2v\cot(\theta)}{\varrho^2}\right) \tag{1.47}$$

$$\frac{\partial v}{\partial t} + u\frac{\partial v}{\partial\varrho} + \frac{v}{\varrho}\frac{\partial v}{\partial\theta} + \frac{uv}{\varrho} - \frac{w^2\cot(\theta)}{\varrho} = -\frac{1}{\rho\varrho}\frac{\partial p}{\partial\theta} + v\left(\nabla^2 v - \frac{v}{\varrho^2\sin^2(\theta)} + \frac{2}{\varrho^2}\frac{\partial u}{\partial\theta}\right) \tag{1.48}$$

$$\frac{\partial w}{\partial t} + u\frac{\partial w}{\partial\varrho} + \frac{v}{\varrho}\frac{\partial w}{\partial\theta} + \frac{uw}{\varrho} - \frac{v\,w\cot(\theta)}{\varrho} = -\frac{1}{\rho\varrho\sin(\theta)}\frac{\partial p}{\partial\varphi} + v\left(\nabla^2 w - \frac{w}{\varrho^2\sin^2(\theta)}\right) \tag{1.49}$$

Here

$$\nabla^2 = \frac{\partial^2}{\partial\varrho^2} + \frac{2}{\varrho}\frac{\partial}{\partial\varrho} + \frac{1}{\varrho^2}\frac{\partial^2}{\partial\theta^2} + \frac{\cot(\theta)}{\varrho^2}\frac{\partial}{\partial\theta} \tag{1.50}$$

The strain rate is

$$d_{\varrho\theta} = \frac{1}{2}\left(\frac{1}{\varrho}\frac{\partial u}{\partial\theta} + \frac{\partial v}{\partial\varrho} - \frac{v}{\varrho}\right) \tag{1.51}$$

Define a stream function

$$u = \frac{1}{\varrho^2\sin(\theta)}\frac{\partial\psi}{\partial\theta}, \quad v = \frac{-1}{\varrho\sin(\theta)}\frac{\partial\psi}{\partial\varrho} \tag{1.52}$$

Then the N–S equation becomes

$$\frac{\partial}{\partial t}E^2\psi + \frac{1}{\varrho^2\sin(\theta)}\frac{\partial(E^2\psi,\psi)}{\partial(\varrho,\theta)} + \frac{2E^2\psi}{\varrho^2\sin^2(\theta)}\left[\cos(\theta)\frac{\partial\psi}{\partial\varrho} - \frac{\sin(\theta)}{\varrho}\frac{\partial\psi}{\partial\theta}\right]$$
$$+ \frac{2\chi}{\varrho^2\sin^2(\theta)}\left[\cos(\theta)\frac{\partial\chi}{\partial\varrho} - \frac{\sin(\theta)}{\varrho}\frac{\partial\chi}{\partial\theta}\right] = \nu E^4\psi \tag{1.53}$$

$$\frac{\partial\chi}{\partial t} - \frac{1}{\varrho^2\sin(\theta)}\frac{\partial(\psi,\chi)}{\partial(\varrho,\theta)} = \nu E^2\chi \tag{1.54}$$

where

$$E^2 = \frac{\partial^2}{\partial\varrho^2} + \frac{1}{\varrho^2}\frac{\partial^2}{\partial\theta^2} - \frac{\cot(\theta)}{\varrho^2}\frac{\partial}{\partial\theta}, \quad \chi = \varrho\sin(\theta)w \tag{1.55}$$

Aside from the well-known Cartesian, cylindrical and spherical coordinate systems, there exist other systems in order to suit the problem. Examples include bipolar, elliptic, parabolic, bi-spheric, ellipsoidal, toroidal, etc.

1.3 Parallel Flow in a Duct

We have discussed plane two-dimensional and axisymmetric flow. A non-plane flow which is not fully three dimensional is the parallel flow and the axisymmetric concentric flow in a duct. In Cartesian coordinates, let $w(x, y)$ denote the velocity in the z direction. The N–S equation degenerates to the equation

$$\frac{\partial w}{\partial t} = -\frac{1}{\rho}\frac{\partial p}{\partial z} + \nu\left(\frac{\partial^2 w}{\partial x^2} + \frac{\partial^2 w}{\partial y^2}\right) + f_z \tag{1.56}$$

In general orthogonal coordinates, it is

$$\frac{\partial w}{\partial t} = -\frac{1}{\rho}\frac{\partial p}{\partial z} + \nu\nabla^2 w + f_z \tag{1.57}$$

Notice the nonlinear momentum terms are absent.

Exercises

(1.1) Deduce the N–S equation which is applicable to the pulsatile (steady and oscillatory) flow through a long circular tube.

(1.2) Figure 1.2 shows Dean's curved tube coordinates where

Fig. 1.2 Curved tube
coordinate system

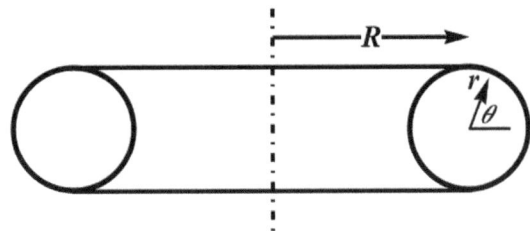

$$x = [R + r\cos(\varphi)]\cos(\theta), \quad y = [R + r\cos(\varphi)]\sin(\theta), \quad z = r\sin(\varphi) \qquad (1.58)$$

Write the Laplace equation $\nabla^2 T = 0$ in (r, φ, θ) coordinates.

(1.3) Write out the N–S equation for a rectangular duct with constant centerline curvature.

Notes

A more detailed derivation of the N–S equation is in Aris (1962). The N–S equation for cylindrical and spherical coordinates in terms of stream function can be found in Langlois and Deville (2014). The N–S equation in non-orthogonal tensor form is in Sokolnikoff (1964). The N–S equation for the flow in a helical tube was given by Wang (1981). Curvilinear coordinates and vector fields are discussed in Morse and Feshbach (1953). The N–S equation for orthogonal coordinates is well discussed in Whitham (1963) and Batchelor (2000).

References

R. Aris, *Vectors, Tensors, and the Basic Equations of Fluid Mechanics* (Prentice-Hall, NJ, 1962)

G.K. Batchelor, *An Introduction to Fluid Dynamics*, Appendix 2 (Cambridge University Press, UK, 2000)

W.E. Langlois, M.O. Deville, *Slow Viscous Flow*, 2nd Ed, Chapter 3, (Springer, NY, 2014)

P.M. Morse, H. Feshbach, *Methods of Theoretical Physics*, Chapter 1 (McGraw-Hill, NY, 1953)

I.S. Sokolnikoff, *Tensor Analysis: Theory and Applications to Geometry and Mechanics of Continua*, 2nd Ed. (Wiley, NY 1964)

C.Y. Wang, J. Fluid Mech. **108**, 185–194 (1981)

G.B. Whitham, in *Laminar Boundary Layers*, ed. by L. Rosenhead (Clarendon, Oxford, 1963)

Exact Solutions

2

The Navier–Stokes (N-S) equation, basically nonlinear partial differential equations, has few analytic exact solutions. Aside from boundary conditions, the N-S equation has two fluid descriptors. Although the constant density can be absorbed into the pressure, the kinematic viscosity (the Reynolds number in non-dimensional form) is a pertinent parameter. By an exact solution we mean an analytic solution to the N-S equation, where the viscosity is arbitrary, is exact for all space and time. Evidently, all closed-form solutions are exact solutions. Numerical solutions and series solutions are not exact. Similarity solutions, where the partial differential equations transform to an ordinary differential equation and resulting in a universal curve with implicit viscosity, are exact solutions. The degenerate potential flow solutions, although satisfying the N-S equation, are excluded.

Exact solutions are important in two aspects. Firstly, they represent fundamental fluid problems, especially those with nonlinear properties. Secondly, they serve as accuracy standards for non-exact solutions, such as those from numerical, asymptotic or experimental means.

There are three types of exact solutions. The first type describes parallel and related flows. In these cases the N-S equation becomes linear due to geometry. The second type are similarity solutions. The third type involves forced linearization (not approximation) of the N-S equation. Usually no solid boundaries can be included.

In this chapter we shall present some examples of exact solutions. They are chosen for their importance in describing the method or the different basic fluid flows. Variations and extensions to these basic solutions such as different boundary geometry, different boundary condition, superposition, etc., are not presented. Comprehensive reviews can be found in Wang (1989, 1991).

C. Y. Wang, *Essential Analytic Laminar Flow*, Synthesis Lectures on Engineering, Science, and Technology, https://doi.org/10.1007/978-3-031-36449-5_2

2.1   Linear Flows

The streamlines are parallel or concentric in these exact solutions. These include the basic Poiseuille flow in a circular tube and the Couette flow due to the relative parallel motion of two plates. Related solutions include the steady flow in an elliptic tube, in an equilateral triangular duct, and the relative motion (translation or rotation) of annular ducts etc. These problems are well discussed in introductory texts and will not be repeated here.

2.1.1   Steady Flow Through a Partially Collapsed Tube

The partially collapsed tube cannot be described in any separable coordinate system. The steady N-S equation in cylindrical coordinates Eq. (1.57) is

$$\frac{\partial^2 w}{\partial r^2} + \frac{1}{r}\frac{\partial w}{\partial r} + \frac{1}{r^2}\frac{\partial^2 w}{\partial \theta^2} = \frac{1}{\rho}\frac{dp}{dz} = -C \tag{2.1}$$

where C represents the negative pressure gradient. Using a torsion analogy Wang (2017) found a family of two-fold symmetric solutions

$$\frac{w}{C} = -\frac{1}{4}r^2 + c_0 + c_2 r^2 \cos(2\theta) + c_4 r^4 \cos(4\theta) \tag{2.2}$$

By adjusting the coefficients, the $w = 0$ curve (the shape of the tube) is obtained a posteriori. Figure 2.1 shows a cross section. Notice the velocity maxima are not at the centroid but closer to the sides.

2.1.2   Oscillating Plate

This is also known as Stokes' second problem. Let the plate be at $y = 0$ and oscillates in its own plane with velocity $U\cos(\omega t)$ where ω is the frequency. For parallel flow, Eq. (1.16) simplifies to

$$\frac{\partial u}{\partial t} = \nu \frac{\partial^2 u}{\partial y^2} \tag{2.3}$$

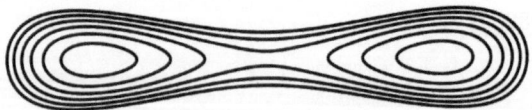

Fig. 2.1 Cross section of a partially collapsed tube showing constant velocity curves

To separate variables in oscillatory flow, let

$$u = f(y)e^{i\omega t} \tag{2.4}$$

where $i = \sqrt{-1}$ and only the real part of the product has physical meaning. Equation (2.3) gives

$$i\omega f = \nu f'' \tag{2.5}$$

Together with the boundary conditions

$$f(0) = U, \quad f(\infty) = 0 \tag{2.6}$$

The solution is

$$f = U e^{-(1+i)\sqrt{\omega/2\nu}\, y} \tag{2.7}$$

Thus

$$u = U e^{-\sqrt{\frac{\omega}{2\nu}}\, y} \cos\left(\omega t - \sqrt{\frac{\omega}{2\nu}}\, y\right) \tag{2.8}$$

Notice, due to the exponential decay, there is a boundary layer of order $\sqrt{\nu/\omega}$ thickness (Stokes' layer) where most of the motion takes place. There is also a phase lag (in time) which varies with the distance from the plate. Figure 2.2 shows some velocity profiles at various times.

Fig. 2.2 Instantaneous velocity profiles for an oscillating plate in a fluid

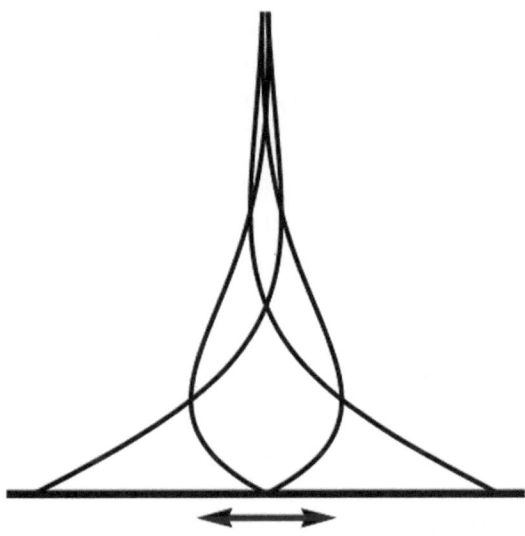

2.1.3 An Infinite Plate Moving in a Rotating System

Motion in a rotating system is important in geophysics (Ekman 1905). The N-S equation with Coriolis force $f = 2\Omega \times u$ is

$$\nu\frac{\partial^2 u}{\partial z^2} + 2\Omega v = 0, \quad \nu\frac{\partial^2 v}{\partial z^2} - 2\Omega u = 0 \tag{2.9}$$

Here, z is the direction perpendicular to the plate which moves with velocity U in the x direction. Let

$$\phi = u + iv \tag{2.10}$$

Equations (2.9) become

$$\nu\frac{\partial^2 \phi}{\partial z^2} - 2i\Omega\phi = 0 \tag{2.11}$$

The solution considering the boundary conditions is

$$\phi = Ue^{-\sqrt{\frac{\Omega}{\nu}}(1+i)z} = u + iv \tag{2.12}$$

Notice the lateral velocity v induced by the Coriolis force.

2.1.4 Impulsively Started Plate

A plate suddenly starts with velocity U in its own plane. The is called Stokes first problem or Rayleigh's problem. The governing equation is still Eq. (2.3), where the initial condition and boundary conditions are

$$u(0, t) = \begin{cases} 0, & t \le 0 \\ U, & t > 0 \end{cases} \tag{2.13}$$

$$u(\infty, t) = 0, \tag{2.14}$$

Notice the method of separation of variables is not applicable. Although Laplace transform is possible for linear sudden-start problems, it is easier to use the method of stretching shown in Example A.1. The similarity equation is

$$2\nu u''(\eta) + \eta u' = 0, \quad \eta = \frac{y}{\sqrt{t}} \tag{2.15}$$

The solution is

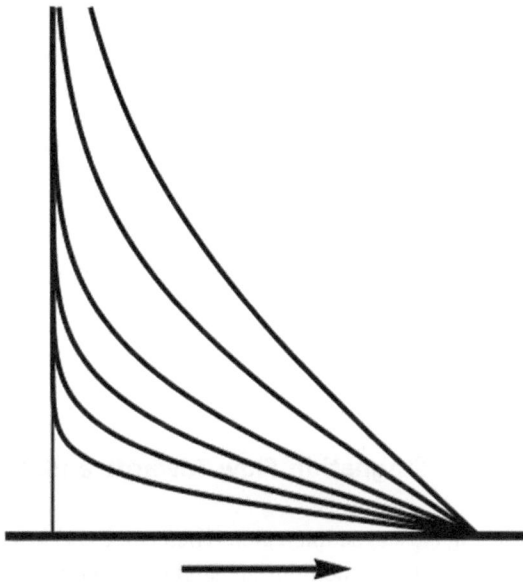

Fig. 2.3 Velocity profiles for an impulsively started plate. Velocity increases as time increases

$$u = U \, \text{erfc}\left(\frac{y}{2\sqrt{vt}}\right) \tag{2.16}$$

where erfc is the complementary error function. Notice the combination of $y/\sqrt{t}$ is typical of unsteady similarity solutions. Figure 2.3 shows the velocity profiles at various times.

2.2 Nonlinear Flows

The exact solutions of all similarity solutions can be obtained from the modified stretching method presented in Appendix A. Representative examples are as follows.

2.2.1 Flow Due to a Stretching Sheet

The flow caused by a stretching sheet occurs in extrusion problems. The boundary conditions are that the velocities

$$u(0) = Ax, \quad v(0) = 0, \quad u(\infty) = 0 \tag{2.17}$$

where A is a measure of stretching velocity. Using Eq. (A.12)

$$\psi = xf(y) \tag{2.18}$$

and substituting into Eq. (1.21) yield the similarity equation

$$f'f'' - ff''' = v f''''$$ (2.19)

The boundary conditions in terms of stream function are

$$f'(0) = A, \quad f(0) = 0, \quad f'(\infty) = 0$$ (2.20)

Crane (1970) found the rare closed-form solution

$$f = \sqrt{Av}\left(1 - e^{-\sqrt{A/v}\,y}\right)$$ (2.21)

2.2.2 Stagnation Flow Towards a Plate

Flow in the stagnation region (Hiemenz 1911) has the same similarity equation, Eq. (2.19). The boundary conditions are slightly different

$$u(0) = 0, \quad v(0) = 0, \quad u(\infty) = Ax$$ (2.22)

However, there is no closed-form solution. In order to obtain a universal similarity solution we transform Eqs. (2.19) and (2.22) such that A and v are absent. Let

$$f = \sqrt{Av}g(\zeta), \quad \zeta = \sqrt{A/v}\,y$$ (2.23)

Then

$$g'g'' - gg''' = g''''$$ (2.24)

$$g'(0) = 0, \quad g(0) = 0, \quad g'(\infty) = 1$$ (2.25)

Using one parameter shooting with a desk computer, we obtain the initial value (and thus the solution)

$$g''(0) = 1.232588$$ (2.26)

Figure 2.4 shows the streamlines. Notice that $A > 0$ or the flow direction cannot be reversed.

Fig. 2.4 Stagnation flow towards a plate

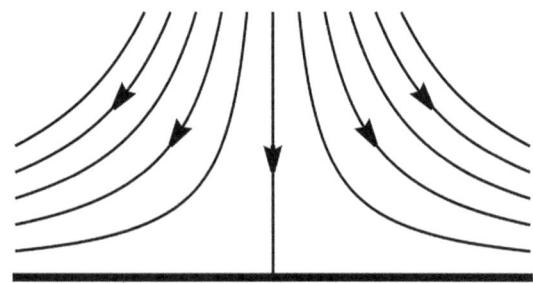

2.2.3 Unsteady Stagnation Flow

The exact solution for nonlinear unsteady flow has the property that three independent variables are combined into one. The basic flow was found by Yang (1958). A transform similar to Eq. (A.16) is

$$\psi = \sqrt{\frac{Av}{(1-\alpha t)}}\, x f(\eta), \quad \eta = \sqrt{\frac{A}{v(1-\alpha t)}}\, y \tag{2.27}$$

The similarity equation is

$$f'''' + f f''' - f' f'' = \frac{S}{2}\left(\eta f''' + 3 f''\right) \tag{2.28}$$

where $S = \alpha/A$ is a non-dimensional parameter representing the importance of unsteadiness. The flow accelerates when S is positive and decelerates when S is negative. In the latter case vorticity is transported into the region, and velocity oscillations may occur.

2.2.4 Flow into a Converging Channel

The flow in a converging channel is also known as Jeffrey-Hamel flow. Equation (A.11) shows all streamlines are radial rays. The boundary of a converging channel suggests using cylindrical coordinates. Let

$$\psi = v f(\theta) \tag{2.29}$$

Equation (1.36) gives the similarity equation

$$f'''' + 2 f f'' + 4 f'' = 0 \tag{2.30}$$

The solution can be expressed in elliptic integrals. Backwards flow may occur if the channel is divergent.

2.2.5   Rotating Disk

Fluid is thrown out by the centrifugal forces of a rotating disk. This problem can be expressed in terms of a stream function and a swirl function. In the spirit of Eq. (A.12) Von Karman (1921) used the primary variables. Let (u, v, w) be velocities in cylindrical coordinates (r, θ, z) respectively. By setting

$$u = \Omega r f'(\eta), \quad v = \Omega r g(\eta), \quad w = -2\sqrt{\Omega v} f(\eta), \quad \eta = \sqrt{\Omega/v} z \tag{2.31}$$

The N-S equation reduces to

$$f''' + 2ff'' - (f')^2 + g^2 = 0, \quad g'' - 2f'g + 2fg' = 0 \tag{2.32}$$

with the boundary conditions

$$f(0) = 0, \quad f'(0) = 0, \quad g(0) = 1, \quad f'(\infty) = 0, \quad g(\infty) = 0 \tag{2.33}$$

The universal curves are integrated numerically. The initial values of the solution are

$$f''(0) = 0.510225, \quad g'(0) = -0.615917 \tag{2.34}$$

2.2.6   Axisymmetric Momentum Jet

The exact solution of a jet of high momentum in an infinite fluid was given by Landau. Use spherical coordinates $(\varsigma, \theta, \phi)$ where ς is the radial distance from the origin, θ is the angle between the ς vector to the z axis. Equation (1.53) gives the N-S equation in the axisymmetric stream function (independent of ϕ). Seeking similarity solution as in Appendix A, we find the only acceptable form is for the stream function be proportional to ς. Let

$$\psi = v\varsigma f(\xi), \quad \xi = \cos(\theta) \tag{2.35}$$

Equation (1.53) reduces to the similarity equation

$$\left(1 - \xi^2\right) f'''' - 4\xi f''' - ff''' - 3f'f'' = 0 \tag{2.36}$$

Landau (1944) gave a rare closed-form solution

$$f = \frac{2(1 - \xi^2)}{(A - \xi)} \tag{2.37}$$

Fig. 2.5 The momentum jet is
indicated by the arrow

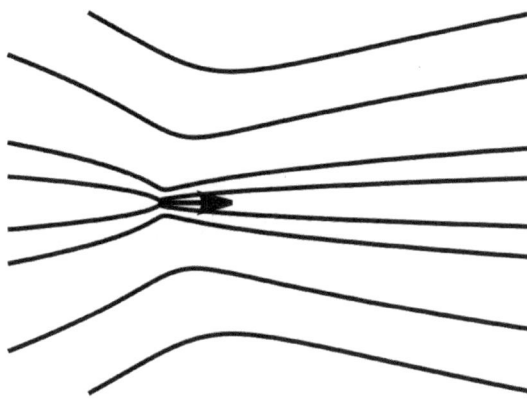

where $1/A$ is related to the strength of the jet. Equation (1.52) shows the velocity is infinite at the source of the jet while the momentum is finite. Figure 2.5 shows the axisymmetric streamlines.

2.3   Beltrami and Related Flows

These flows are obtained by forced linearization. The idea is to set the linear terms and the nonlinear terms in the N-S equation separately to zero, and seek a solution that satisfies both. Usually, the solution is rotational and cannot admit any solid boundaries. More significant examples are presented here.

2.3.1   Source or Vortex in Shear Flow

Tsien (1943) found that a two-dimensional source (or vortex) in a constant shear flow which satisfies the N-S equation.

$$\psi = ay + by^2 + c\tan^{-1}\left(\frac{y}{x}\right) \tag{2.38}$$

$$\psi = ay + by^2 + c\ln(x^2 + y^2) \tag{2.39}$$

Here a, b, c are constants. See Fig. 2.6.

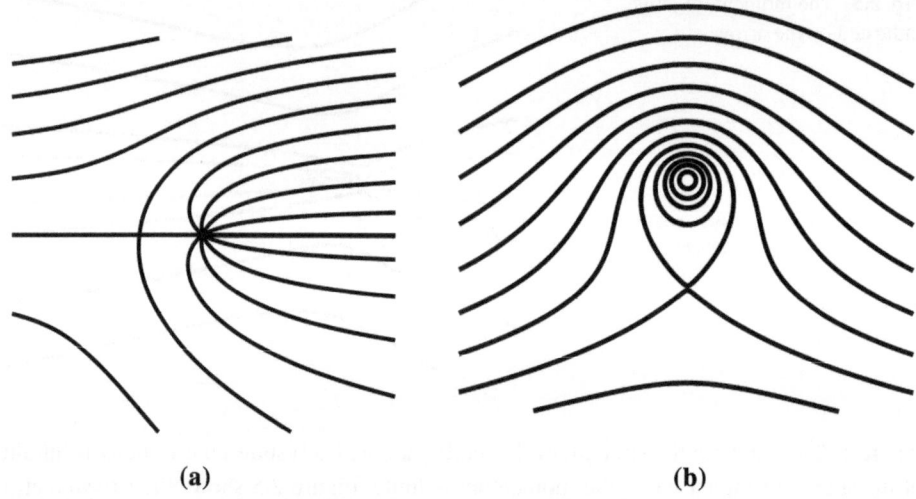

(a)   (b)

Fig. 2.6 **a** A source in shear flow **b** a vortex in shear flow

2.3.2 Ellipsoidal Vortex

In axisymmetric cylindrical coordinates, the internal rotation inside an ellipsoidal bubble is represented by

$$\psi = ar^2\left(\frac{r^2}{b^2} + \frac{z^2}{c^2} - 1\right) \tag{2.40}$$

This is an extension of Hill's (1894) spherical vortex. Figure 2.7 shows an axisymmetric oblate ellipsoidal bubble.

2.3.3 Decay of an Array of Vortices

Consider a doubly infinite rectangular array of vortices which decays in time

$$\psi = a\cos(mx)\cos(ny)e^{-\nu(m^2+n^2)t} \tag{2.41}$$

This is an extension of Taylor's (1923) decay of square vortices. Notice the smaller the vortices the faster the decay.

Fig. 2.7 The ellipsoidal
bubble with internal circulation

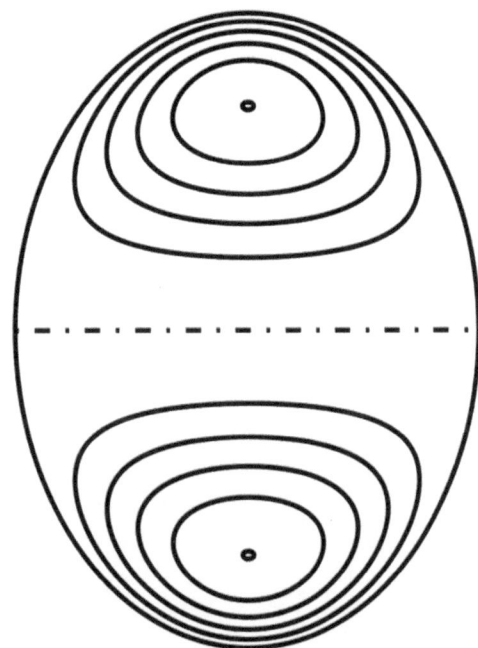

2.3.4 Decay of Flow Downstream of a Row of Cylinders

The modelling of a wake behind a cylinder has always been difficult. Consider the symmetric wake behind a row of cylinders normal to the flow with decaying velocity. The flow near the back of the recirculating wake may be modeled by

$$\psi = \frac{1}{1 + ae^{vht}}\left[Vy - b\sin(my)e^{(-vhx/V)}\right] \tag{2.42}$$

Here

$$h = \frac{mV}{v}, \quad \text{or} \quad h = \frac{\sqrt{V^4 + 4v^2m^2V^2} - V^2}{2v^2} \tag{2.43}$$

This solution is an extension of the steady wake found by Kovasznay (1948). Figure 2.8 shows a periodic wake emanating from the back of cylinders. The streamlines decay in time.

Exercises

(2.1) Find the exact solution to the steady flow in an (a) equilateral triangular duct of side length L (b) elliptic duct of chord length L. You may consult other texts or the internet.

Fig. 2.8 Streamlines in the
back of a row of cylinders

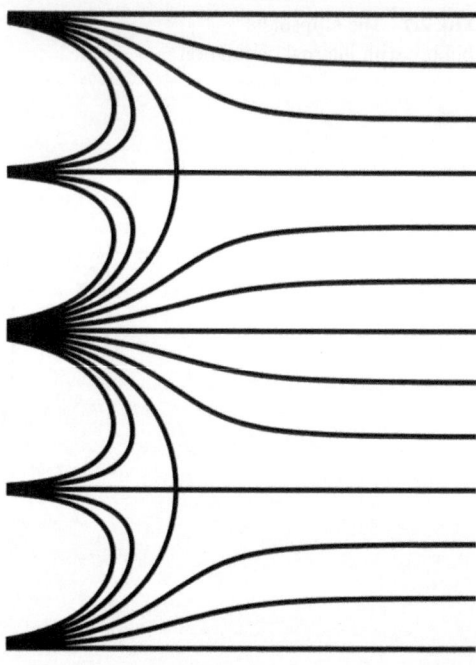

(2.2) Find the solution to the flow in a three-fold partially collapsed tube similar to the
two-fold collapse of Sect. 2.1.1.

(2.3) Consider the flow in a channel due to an oscillatory pressure gradient. What are the
characteristics of the velocity for (a) low frequency (b) high frequency?

(2.4) Find the similarity transform for the axisymmetric stretching of a surface. Then
integrate numerically for the universal curve.

(2.5) Use the Science Citation Index and find all the references to "similarity" and
"unsteady rotation" and "disk or disc" and "Navier–Stokes".

Notes

Unsteady exact N-S solutions were reviewed in Wang (1989). The steady state exact N-S
solutions were reviewed in Wang (1991). Backwards solution for the flow in a collapse
tube was given in Wang (2017). Ekman's original work on ocean currents was in Ekman
(1905). Flow due to a stretching sheet was found by Crane (1970) and reviewed in Wang
(2011). The basic steady stagnation flow was given by Hiemenz (1911) and reviewed in
Wang (2008). The basic unsteady stagnation flow was found by Yang (1958). The flow into
a converging channel was found independently by Jeffrey (1915) and Hamel (1916) and
discussed by Whitham (1963). The rotating disk solution was found by von Karman (1921).
The momentum jet solution was given by Landau (1944) and Landau and Lifshitz (1959).
The references for Sect. 2.3 are Tsien (1943), Hill (1894), Taylor (1923), Kovasznay (1948).
The reference for an axisymmetric stretching surface is in Wang (1984).

References

L.J. Crane, ZAMP **21**, 645–647 (1970)

V.W. Ekman, Ark. Mat. Astron. Fys. **2**, 1–52 (1905)

G. Hamel, Jahresher. Dtsch. Mat. Ver. **25**, 34–60 (1916)

K. Hiemenz, Dinglers Polytech. J. **326**, 321–324 (1911)

M.J.M. Hill, Phil. Trans. R. Soc. London A **185**, 213–245 (1894)

G.B. Jeffrey, Phil Mag. Ser. 6, **29**, 455–465 (1915)

L.I. Kovasznay, Proc. Camb. Phil. Soc. **44**, 58–62 (1948)

L.D. Landau, Dokl. Akad. Nauk. SSSR **43**, 286–288 (1944)

L.D. Landau and E.M. Lifshitz, *Fluid Mechanics* (Pergamon, London, 1959)

H.S. Tsien, Quart. Appl. Math. **1**, 130–148 (1943)

G.I. Taylor, Phil. Mag. Ser. 6 **46**, 671–674 (1923)

T. Von Karman, ZAMM **1**, 233–252 (1921)

C.Y. Wang, Phys. Fluids **27**, 1915–1917 (1984)

C.Y. Wang, Appl. Mech. Rev. **42**(11 Part 2), S269–S282 (1989)

C.Y. Wang, Ann. Rev. Fluid Mech. **23**, 159–177 (1991)

C.Y. Wang, Europ. J. Mech. B/Fluids **30**, 475–479 (2011)

C.Y. Wang, Mech. Res. Comm. **83**, 65–67 (2017)

C.Y. Wang, Europ. J. Mech. B/Fluids **27**, 678–683 (2008)

G.B. Whitham, in *Laminar Boundary Layers*, ed. by L. Rosenhead (Clarendon, Oxford, 1963)

K.T. Yang, J. Appl. Mech. **25**, 421–427 (1958)

Non-dimensionalization, Scaling and Approximations

3

3.1 Dimensions and the Pi Theorem

Every variable or constant can be classified by its dimension. The fundamental dimensions are mass [M], length [L], time [T], temperature $[\theta]$, electric charge $[q]$, etc. The derived dimensions include velocity [L/T], force $[ML/T^2]$, density $[M/L^3]$, kinematic viscosity $[L^2/T]$, etc. But physical truths could not have (man-made) dimensions and the variables must appear in nondimensional groups with dimension one or [1].

The Pi theorem states.

- if there are n parameters and k fundamental dimensions, then there are $n\text{-}k$ non-dimensional groups (π_i). The non-dimensional groups are functionally related (proof omit).

For example, consider a hanging pendulum. The possible parameters governing its motion could include the mass of the hanging mass m [M], the length of the string l [L], the gravitational acceleration g $[L/T^2]$, the initial angle of the string to the vertical ϕ_0 [1] and the present angle ϕ [1] and the present time t [T]. There are $n = 6$ parameters (m, l, g, ϕ_0, ϕ, t) and $k = 3$ fundamental dimensions (M, L, T), so we can form $6\text{--}3 = 3$ non-dimensional groups.

They are $\pi_1 = \phi, \pi_2 = \phi_0, \pi_3 = \sqrt{g/l}\,t$. The functional relation is

$$\pi_1 = f(\pi_2, \pi_3) \quad \text{or} \quad \phi = f\left(\phi_0, \sqrt{\frac{g}{l}}\,t\right) \tag{3.1}$$

Notice that the mass m is not a factor. For the period P of the pendulum, set $t = P$ and $\phi = \phi_0$. Inverting from Eq. (3.1b) we find

$$P = \sqrt{\frac{l}{g}} h(\phi_0) \tag{3.2}$$

where h is a function only of the initial angle. Thus, we found the period is proportional to $l^{1/2}$ and $g^{-1/2}$ without solving any equation.

Some guidelines on finding the non-dimensional groups:

- Include all the possible factors which may affect the outcome.
- Group as simple as possible.
- Keep obscure parameters isolated.

3.2   Non-dimensionalization and Scaling of the N-S Equation

Scaling means all such Pi groups should be of magnitude order one, except perhaps for some constant non-dimensional groups. Magnitude order one of a non-dimensional ratio means, loosely, in the interval 0.1 to 10. This differs from asymptotic order which is defined in Appendix B.

Consider the N-S equation in two-dimensional Cartesian coordinates, Eq. (1.21)

$$\frac{\partial}{\partial t^*} \nabla^{*2} \psi^* + \frac{\partial(\nabla^{*2} \psi^*, \psi^*)}{\partial(x^*, y^*)} = \nu \nabla^{*4} \psi^* \tag{3.3}$$

Here the asterisk denotes dimensional quantity. The initial and boundary conditions may contain some time scale T in t^*, some length scale L in x^* and y^*, and some velocity scale U in $\nabla^* \psi^*$. Scale the dimensional variables so that they are of order one

$$t = \frac{t^*}{T}, \quad x = \frac{x^*}{L}, \quad y = \frac{y^*}{L}, \quad \psi = \frac{\psi^*}{UL} \tag{3.4}$$

Keep in mind that the problem may have several length, velocity, or time scales. The appropriate normalization would give the right estimate of the variable's magnitude.

The N-S equation in non-dimensional form becomes

$$S \frac{\partial}{\partial t} \nabla^2 \psi + \frac{\partial(\nabla^2 \psi, \psi)}{\partial(x, y)} = \frac{1}{R} \nabla^4 \psi \tag{3.5}$$

where S is the Strouhal number representing the importance of unsteadiness (frequency) to momentum and R is the Reynolds number representing the importance of momentum to viscous effects.

$$S = \frac{L}{UT}, \quad R = \frac{UL}{\nu} \tag{3.6}$$

Now all variables in Eq. (3.5) are non-dimensional and scaled to order one. Both S and R are non-dimensional, but their magnitudes may not be of order one, leading to different approximations.

3.3   Approximations of the N-S Equation

In most cases there is no exact solution to the N-S equation. Approximations usually simplify the equation and yield non-exact solutions, which may still capture the fluid dynamic phenomena. We shall illustrate with the two-dimensional Cartesian equation. Approximations in other coordinate systems are similar.

3.3.1   Steady Flow

If the problem is steady, put $S = 0$. If the problem only vary slowly with time, or S is very small, we can still approximate by the steady-state equation (quasi-steady) with an error of order $O(S)$.

3.3.1.1   Steady Flow and Small Reynolds Number

The steady-state equation has the Reynolds number R. If R is small the N-S equation is approximated by the Stokes' equation. In two-dimensional Cartesian coordinates it is

$$\nabla^4 \psi = 0 \tag{3.7}$$

The Stokes equation contains the highest partial derivatives. Its solution is discussed in Chap. 6. The error of Stokes flow is $O(R)$.

3.3.1.2   Steady Flow and Large Reynolds Number

For steady flow when R is large (but still remain laminar), Eq. (3.5) gives the approximation

$$\frac{\partial\left(\nabla^2 \psi, \psi\right)}{\partial(x, y)} = 0 \tag{3.8}$$

Mathematically, Eq. (3.8) indicates the functional relationship

$$\nabla^2 \psi = F(\psi) \tag{3.9}$$

which is related to Beltrami flow. Physically, Eq. (3.9) means the vorticity follows the streamlines, or non-diffusive inviscid flow. For uniform flow originated at infinity, Eq. (3.9) shows the vorticity is zero and remains zero. Thus, Eq. (3.8) reduces to inviscid irrotational flow or potential flow

$$\nabla^2 \psi = 0 \tag{3.10}$$

In Appendix C, we shall discuss more about potential flow and the velocity potential.

However, the potential flow equation is only a second order partial differential equation, and cannot satisfy all the boundary conditions, especially a no-slip condition. Since the N-S equation is singular for large R (degenerates to a lower order differential equation) there exists a boundary layer near the solid boundary to satisfy the no-slip condition. See Chap. 4 for more details on the boundary layer theory. The resulting solution has an error of $O(1/R)$. Higher order corrections can be obtained using matched asymptotic expansions in Appendix B.

3.3.2   Unsteady Flow and $S \gg 1$

For large S, the unsteady term is more important than the nonlinear momentum terms. These cases occur in high frequency oscillations or sudden start problems. There may be unsteady boundary layers locally similar to Stokes first and second problems (Sects. 2.1.2 and 2.1.4). Unsteady problems are discussed in Chap. 5.

3.3.2.1 $S \gg 1$ and $R \geq O(1)$
The leading order is

$$\frac{\partial}{\partial t} \nabla^2 \psi = 0 \tag{3.11}$$

This means the vorticity does not change with time. Thus if the flow is potential at some time, it will be potential thereafter.

However, the degree in space variables is decreased. So there will be a boundary layer on the solid surface. The boundary layer thickness is of order $(SR)^{-1/2}$.

3.3.2.2 $S \gg 1$ and $R \ll 1$

(a) If $SR = O(1)$, then the approximation is

$$SR\frac{\partial}{\partial t} \nabla^2 \psi = \nabla^4 \psi \tag{3.12}$$

or unsteady Stokes flow.

(b) If $SR \ll 1$,

It is quasi-steady Stokes flow governed by Eq. (3.7).

3.3.3 Flow in a Thin Layer

Another approximation is the flow in a thin layer. Applications include the flow in a boundary layer, film flow, forced flow in crevices or in lubrication problems.

Let x^* be of order L, but y^* is much smaller, of order δL, where $\delta \ll 1$. Since

$$u^* = \frac{\partial \psi^*}{\partial y^*} = O(U) \tag{3.13}$$

the appropriate scaling is

$$t = \frac{t^*}{T}, \quad x = \frac{x^*}{L}, \quad \eta = \frac{y^*}{\delta L}, \quad \psi = \frac{\psi^*}{\delta U L} \tag{3.14}$$

where the non-dimensional variables are of order unity. The leading terms of Eq. (3.5) are

$$\frac{S}{\delta^2} \frac{\partial^3 \psi}{\partial \eta^2 \partial t} + \frac{1}{\delta^3} \frac{\partial(\frac{\partial^2 \psi}{\partial \eta^2}, \psi)}{\partial(x, \eta)} = \frac{1}{R \delta^4} \frac{\partial^4 \psi}{\partial \eta^4} \tag{3.15}$$

The ratio of the three terms are $(S\delta)$, 1, $1/(R\delta)$. The approximations of Eq. (3.15) are much simpler than those of Eq. (3.5).

3.3.3.1 Boundary Layer and Lubrication Equations

If S is zero, or $(S\delta) < < 1$, the flow is steady or quasi-steady. Equation (3.15) becomes

$$\frac{\partial(\frac{\partial^2 \psi}{\partial \eta^2}, \psi)}{\partial(x, \eta)} = \frac{1}{R\delta} \frac{\partial^4 \psi}{\partial \eta^4} \tag{3.16}$$

(a) If $(R\delta) \geq O(1)$, Eq. (3.16) yields the high Reynolds number boundary layer equation.

Equation (3.16) can be integrated with respect to η to obtain

$$\frac{\partial(\frac{\partial \psi}{\partial \eta}, \psi)}{\partial(x, \eta)} = \frac{1}{R\delta} \frac{\partial^3 \psi}{\partial \eta^3} + F(x) \tag{3.17}$$

Since

$$u = \frac{\partial \psi}{\partial \eta}, \quad v = -\frac{\partial \psi}{\partial x} \tag{3.18}$$

Equation (3.17) gives

$$u \frac{\partial u}{\partial x} + v \frac{\partial u}{\partial \eta} = \frac{1}{R\delta} \frac{\partial^2 u}{\partial \eta^2} + F(x) \tag{3.19}$$

Here, $F(x)$ can be identified as the pressure gradient outside the boundary layer, or

$$F(x) = U \frac{dU}{dx} \tag{3.20}$$

(b) If $(R\delta) \ll 1$, the leading order of Eq. (3.16) is

$$\frac{\partial^4 \psi}{\partial \eta^4} = 0 \tag{3.21}$$

This gives the simple lubrication flow approximation (Sect. 6.9.1). The solution is a cubic polynomial in η with coefficients functions of x.

3.3.3.2 Highly Unsteady Flow

If $(S\delta) \gg 1$, the nonlinear momentum terms can be ignored. Equation (3.15) gives

$$\frac{\partial^3 \psi}{\partial \eta^2 \partial t} = \frac{1}{SR\delta^2} \frac{\partial^4 \psi}{\partial \eta^4} \tag{3.22}$$

which is a diffusion equation in vorticity. Whether boundary layers exist or not depends on the order of $(SR\delta^2)$. Such highly unsteady flow in a thin layer occurs in high frequency flow and sudden start problems in a boundary layer. See Chap. 5.

3.4   Stability

Solutions to the N-S equation may be unstable for different reasons.

One reason has to do with the sensitivity to boundary conditions, especially when vorticity spreads throughout the region of interest. For example, the solution to the Von Karman rotating disk problem (Sect. 2.2.5) is fairly stable since vorticity is generated and confined near the disk. However, for Bodewadt (1940) flow (rotating flow over a fixed disk) the vorticity permeates the entire region. The solution is oscillatory (in space) and sensitive to the boundary conditions at infinity. Another example is the unsteady squeeze flow between two plates (Sect. 5.3). When the plates move towards each other, the vorticity is confined, and the solution is smooth and stable. When the plates are pulled apart, the solution is oscillatory and sensitive to the conditions at infinity.

Another source of instability is the transition from laminar to turbulent. There are many causes of the transition. Here we are only interested in the small disturbance theory of viscous flow.

Consider the two-dimensional N-S equation in Cartesian coordinates. Let the stream function be the sum

$$\psi = \Psi(x, y, t) + \tilde{\psi}(x, y, t) \tag{3.23}$$

Here Ψ is the basic N-S solution found and $\tilde{\psi}$ is a small perturbation whose growth would suggest instability. Substitution into Eq. (3.5) gives

$$S\frac{\partial}{\partial t}\nabla^2(\Psi + \tilde{\psi}) + \frac{\partial(\nabla^2(\Psi + \tilde{\psi}), (\Psi + \tilde{\psi}))}{\partial(x, y)} = \frac{1}{R}\nabla^4(\Psi + \tilde{\psi}) \tag{3.24}$$

Now cancel the terms satisfied by Ψ and ignore the smaller products of $\tilde{\psi}$. The linearized approximation is

$$S\frac{\partial}{\partial t}\nabla^2\tilde{\psi} + \frac{\partial\left(\nabla^2\Psi, \tilde{\psi}\right)}{\partial(x, y)} + \frac{\partial(\nabla^2\tilde{\psi}, \Psi)}{\partial(x, y)} = \frac{1}{R}\nabla^4\tilde{\psi} \tag{3.25}$$

3.4.1 The Orr-Sommerfeld Equation

When the basic flow Ψ is a steady parallel flow Eq. (3.25) reduces to

$$S\frac{\partial}{\partial t}\nabla^2\tilde{\psi} - U''(y)\frac{\partial\tilde{\psi}}{\partial x} + U(y)\frac{\partial\nabla^2\tilde{\psi}}{\partial x} = \frac{1}{R}\nabla^4\tilde{\psi} \tag{3.26}$$

Here, $U = d\Psi/dy$. If the perturbation is composed of normal modes let

$$\tilde{\psi} = \phi(y)e^{i\alpha(x-ct)} \tag{3.27}$$

Here, ϕ is a complex amplitude, α is the wave number, and c is the wave speed. The time scale is now $1/(\alpha c)$ and $S = \alpha cL/U$. Equation (3.26) becomes the Orr-Sommerfeld equation

$$(U - cS)(\phi'' - \alpha^2\phi) - U''\phi = \frac{-i}{R\alpha}(\phi'''' - 2\alpha^2\phi'' + \alpha^4\phi) \tag{3.28}$$

This equation has been studied for a variety of parallel flows. We shall leave its solution to the references.

Exercises

(3.1) Consider the eccentric rotation of a circular cylinder in a circular casing. Simplify the N-S equation when the gap width is small compared to the radii.

(3.2) In Exercise (3.1), further assume (a) the fluid is very viscous (b) the rotation speed is very large.

Notes

Dimensional analysis and the Pi theorem are discussed in Birkhoff (1960). Instability and oscillations are found in Bodewadt (1940) and Wang (1976). Introduction to stability of laminar flows is found in Sherman (1990), White (2006) and Leal (2010).

References

G. Birkhoff, *Hydrodynamics-A Study in Logic, Fact and Similitude* (Princeton University, NJ, 1960)

U.T. Bodewadt, ZAMM **20**, 241–253 (1940)

L.G. Leal, *Advanced Transport Phenomena* (Cambridge University UK, 2010)

F.S. Sherman, *Viscous Flow* (McGraw-Hill, NY, 1990)

C.Y. Wang, J. Appl. Mech. **43**, 579–583 (1976)

F.M. White, *Viscous Fluid Flow*, 3rd edn. (McGraw-Hill, MA, 2006)

Boundary Layers

4

The boundary layer equation is an approximation of the N-S equation at high Reynolds numbers. The N-S equation becomes singular (Chap. 3). The leading terms show the outer flow is inviscid (and usually potential). Since the outer flow cannot satisfy the no-slip condition, a boundary layer near a solid surface exists.

We shall start with establishing the steady, two-dimension boundary layer equation and its solutions, and then extend to the axisymmetric cases. The unsteady boundary layer will be considered in Chap. 5.

4.1 The Boundary Layer Equation

Equation (3.16) gives the leading terms of a boundary layer near the surface $y = 0$. In dimensional form they are

$$\frac{\partial(\psi_{yy}, \psi)}{\partial(x, y)} = \nu \psi_{yyyy} \tag{4.1}$$

The outer flow stream function satisfies the potential equation

$$\Psi_{xx} + \Psi_{yy} = 0 \tag{4.2}$$

Equation (4.1) can be integrated once

$$\psi_y \psi_{yx} - \psi_x \psi_{yy} = F(x) + \nu \psi_{yyy} \tag{4.3}$$

where $F(x)$ is related to the streamwise pressure. The boundary layer equation in terms of the stream function, Eq. (4.3), is more advantageous than using primary variables because

© The Author(s), under exclusive license to Springer Nature Switzerland AG 2024
C. Y. Wang, *Essential Analytic Laminar Flow*, Synthesis Lectures on Engineering,
Science, and Technology, https://doi.org/10.1007/978-3-031-36449-5_4

(1) continuity equation is eliminated, (2) pressure across the boundary layer can be shown to be constant, and (3) it is easier for similarity transforms.

Since the inner ψ must match the outer Ψ, and the latter is zero on the boundary, Eq. (4.3) shows

$$F(x) = \Psi_y \Psi_{yx} = U(x)U'(x) = -\frac{1}{\rho}\frac{dP}{dx} \tag{4.4}$$

where P is the streamwise pressure of potential flow. In terms of velocities, Eq. (4.3) is the more familiar

$$uu_x + vu_y = UU'(x) + vu_{yy} \tag{4.5}$$

The boundary conditions to Eq. (4.3) are no slip for the inner solution

$$\psi(0) = 0, \quad \psi_y(0) = 0 \tag{4.6}$$

and the inner flow matches the outer flow

$$\psi_y(x, \infty) \sim \Psi_y(0) = U(x) \tag{4.7}$$

Equations (4.5)–(4.7) also applies to two-dimensional curved boundaries, provided the radius of curvature is of order larger than the boundary layer thickness, and thus, the curvature terms do not enter the boundary layer equation. We may regard (x, y) as intrinsic coordinates, with the x-axis coinciding with the solid surface and the y axis normal to the surface.

4.1.1 Axisymmetric Boundary Layer Equation

The axisymmetric cylindrical coordinate system (r, z) can be used. Consider a boundary layer on an axisymmetric surface $r = r_0(z)$. If r_0 is much larger than the boundary layer thickness, Eq. (1.46) reduces to

$$E^2 \sim \frac{\partial^2}{\partial r^2} \tag{4.8}$$

and from Eq. (1.44), the steady boundary layer equation is

$$-\frac{1}{r}\frac{\partial(\psi_{rr}, \psi)}{\partial(r, z)} + \frac{2}{r^2}\psi_{rr}\psi_z = v\psi_{rrrr} \tag{4.9}$$

Now, if r_0 is zero or the same order as the boundary layer thickness, let

$$E^2 = \frac{\partial^2}{\partial r^2} - \frac{1}{r}\frac{\partial}{\partial r} \tag{4.10}$$

Equation (1.44) is

$$-\frac{1}{r}\frac{\partial\left(E^2\psi, \psi\right)}{\partial(r, z)} + \frac{2}{r^2}E^2\psi\psi_z = \nu E^4\psi \tag{4.11}$$

If the boundary layer is along a constant z surface, set

$$E^2 = \frac{\partial^2}{\partial z^2} \tag{4.12}$$

in Eq. (4.11).

4.2   Similarity Solutions

These solutions are exact solutions of the boundary layer equation. A similarity transform (Appendix A) reduces the partial differential equation to an ordinary differential equation which can be easily integrated.

4.2.1   Two-Dimensional Solutions

Substitute Eq. (A.1)

$$\psi = x^m f(\eta), \quad \eta = \frac{y}{x^n} \tag{4.13}$$

into Eq. (4.1). Similarity is possible if

$$m = 1 - n \tag{4.14}$$

resulting in

$$(3m - 2)f'f'' - mff''' = \nu f'''' \tag{4.15}$$

Equation (4.15) can be integrated once

$$(2m - 1)\left(f'\right)^2 - mff'' = \nu f''' + C \tag{4.16}$$

where C is a constant.
    The tangential velocity is

$$u = \frac{\partial\psi}{\partial y} = x^{2m-1}f'(\eta) \tag{4.17}$$

If at the edge of the boundary layer the velocity is given

$$U = x^{2m-1} f'(\infty) = x^\alpha K, \quad \alpha = 2m - 1 \tag{4.18}$$

then

$$f'(\infty) = K \tag{4.19}$$

In order to obtain a universal curve, we need to eliminate the parameters ν and K. Set

$$f(\eta) = A g(\zeta), \quad \zeta = B\eta \tag{4.20}$$

Then adjust A, B so that ν and K do not appear, e.g.

$$A = \sqrt{\nu K}, \quad B = \sqrt{K/\nu} \tag{4.21}$$

The result is

$$(2m - 1)\left[(g')^2 - 1\right] - m g g'' = g''' \tag{4.22}$$

$$g'(\infty) = 1 \tag{4.23}$$

Some similarity solutions are given below.

4.2.1.1 Boundary Layer on a Wedge

This is the Falkner-Skan (1931) boundary layer solution for the flow over a wedge. Figure (4.1) shows a flow towards a wedge with half angle β. The outer flow stream function is proportional to $r^\lambda \sin(\lambda\theta)$ inducing a tangential velocity Eq. (4.18). We find $\lambda = \alpha + 1$. The wedge angle and the velocity power are related by

$$\alpha = \frac{\beta}{\pi - \beta} \tag{4.24}$$

The no-slip boundary conditions for Eq. (4.22) are

Fig. 4.1 Flow over a wedge
with half vertex angle β.
Dashed curves show the extent
of the boundary layer

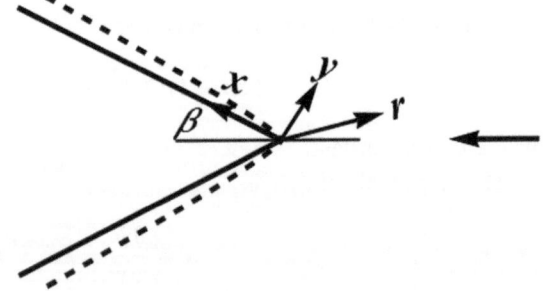

Table 4.1   Some special wedge cases

β	α	m	$g''(0)$	Comment
0.5π	1	1	1.23259	Stagnation flow normal to a plate
0	0	0.5	0.332056	Blasius (1908) flow parallel to a semi-infinite plate
-0.099442π	-0.090429	0.454786	0	Separation occurs for a slight corner expansion

$$g(0) = 0, \quad g'(0) = 0 \tag{4.25}$$

Although the half wedge angle can be any angle, of particular interest are the cases shown in Table 4.1. We shall not discuss the stagnation flow normal to a plate since it has an exact N-S solution which would yield the boundary layer solution when expanded for large Reynolds numbers. Similarly, we shall not discuss the boundary layer for the radial flow into a sink. The boundary layer flow over a semi-infinite plate ($\beta = 0$) is important as it also serves as a benchmark for numerical or experimental accuracy (Blasius 1908).

In general, when β is negative the outer flow is towards the vertex on one surface and away from the vertex on the other surface. Since vorticity is transported into the domain from infinity, these flows may have the properties of non-uniqueness, oscillations (in space), reverse flow and especially boundary layer separation, which destroys the prescribed outer flow.

4.2.1.2   Free Shear Layers

These flows do not have a solid material boundary. We present three typical cases.

(1) Shear layer due to the interaction of two uniform streams.

This flow, due to Lock (1951), is obtained by matching the velocities and the stresses at the interface of the two uniform streams. The formulation is similar to Sect. 4.2.1.1 except for the boundary conditions.
(2) The momentum jet.

Envision the two-dimensional analog of the Landau axisymmetric jet (Sect. 2.2.6). The momentum across any transverse plane is constant

$$M = \rho \int_{-\infty}^{\infty} u^2 dy \tag{4.26}$$

Using Eqs. (4.17, 4.18) we find

$$m = \frac{1}{3}, \quad n = \frac{2}{3}, \quad \alpha = -\frac{1}{3} \tag{4.27}$$

Equation (4.16) yields

$$-\frac{1}{3}(f')^2 - \frac{1}{3}ff'' = \nu f''' \tag{4.28}$$

since there is no pressure gradient at infinity. Equation (4.26) becomes

$$J = \frac{M}{2\rho} = \int_0^\infty (f')^2 d\eta \tag{4.29}$$

Similar to Sect. 4.2.1, we can obtain a universal equation by

$$f(\eta) = (\nu J)^{\frac{1}{3}} g(\zeta), \quad \zeta = \left(\frac{J}{\nu^2}\right)^{1/3} \eta \tag{4.30}$$

Then

$$-\frac{1}{3}\left[(g')^2 + gg''\right] = g''' \tag{4.31}$$

$$g(0) = 0, \quad g''(0) = 0, \quad \int_0^\infty (g\prime)^2 d\zeta = 1 \tag{4.32}$$

Schlichting (1933) found the closed-form solution

$$g = 3^{2/3} \tanh\left(\frac{3^{2/3}\zeta}{6}\right) \tag{4.33}$$

The initial value is $g'(0) = \frac{3^{\frac{1}{3}}}{2} = 0.721125$.

(3) Linear shear flow.

If the outer velocity is a power-law shear flow proportional to y^γ

$$U = y^\gamma h(\eta) = x^{n\gamma} \eta^\gamma h(\eta), \quad h(\infty) = H \tag{4.34}$$

Comparing Eq. (4.34) with Eq. (4.18) we find $f'(\infty) = H\eta^\gamma$ and then Eq. (4.16) cannot be satisfied for any constant C, unless $\gamma = 0$ (wedge flow) or $\gamma = 1$ (linear shear flow). Notice, for linear shear flow, the outer solution already satisfies the N-S equation and the no-slip condition on a solid surface.

However, a boundary layer develops if the boundary is other than solid, as in Lock's two-stream interaction. Wang (1992) studied the linear shear flow over a still fluid.

4.2.1.3 Stretching Boundary

Consider the flow caused by a stretching sheet emanating from a slit into a still fluid. The problem models fluid flows caused by extrusion or surface shear. The velocity of the sheet is given and directed away from the source point.

(1) Power law sheet velocity.

Let the velocity be $u = K x^\alpha$, where K now a measure of the velocity magnitude. Since the flow is quiescent at infinity, the constant C in Eq. (4.16) is zero. The boundary conditions are

$$f(0) = 0, \quad f'(0) = K, \quad f'(\infty) = 0 \qquad (4.35)$$

Using Eqs. (4.20, 4.21), the similarity equation is

$$(2m - 1)(g')^2 - mgg'' = g''' \qquad (4.36)$$

$$g(0) = 0, \quad g'(0) = 1, \quad g'(\infty) = 0 \qquad (4.37)$$

Table 4.2 shows some specific cases.

Although the power law exterior velocity problem and the power law stretching sheet problem share the same Eq. (4.16), some boundary conditions are interchanged. Due to nonlinearity the solutions are not related. Figure 4.2a shows a comparison of Blasius uniform flow over a semi-infinite plate and Sakiadis flow due to pulling a plate from a slit. Similarly, Fig. 4.2b compares the stagnation flow on a plate and Crane's linearly stretching sheet.

(2) Exponential sheet velocity.

Let the velocity of the sheet be exponential, $u = K e^{2ax}$. Let

$$\psi = e^{ax} f(\eta), \quad \eta = e^{ax} y \qquad (4.38)$$

Equation (4.1) reduces to

Table 4.2 Some special stretching sheet cases

α	m	$g''(0)$	Comment
1	1	−1	Crane's (1970) stretching sheet exact solution (Sect. 2.2.1)
0	1/2	−0.443748	Sakiadis (1961) flow due to extrusion of a plate with uniform velocity
−1/3	1/3	0	The 2D momentum jet (Sect. 4.2.1.2)

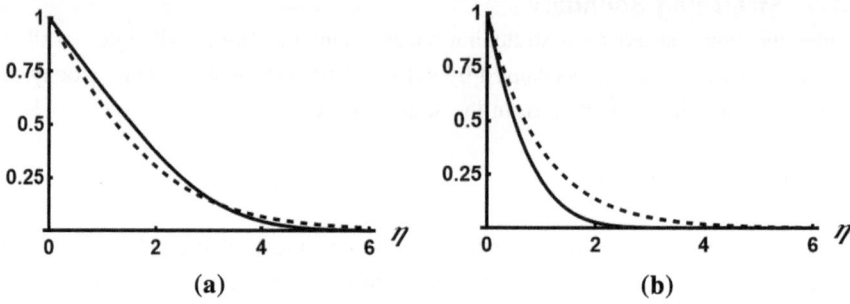

Fig. 4.2 **a** Solid curve: $1 - u$ of Blasius flow; dashed curve: u of Sakiadis flow. **b** Solid curve: $1 - u$ of stagnation flow; dashed curve: u of Crane flow

$$3 f' f'' - f f'' = \frac{v}{a} f'''' \tag{4.39}$$

The boundary conditions are

$$f(0) = 0, \quad f'(0) = K, \quad f'(\infty) = 0 \tag{4.40}$$

Using the transform

$$f(\eta) = \sqrt{\frac{vK}{a}} g(\zeta), \quad \zeta = \sqrt{\frac{aK}{v}} \eta \tag{4.41}$$

and after integrating once, Eq. (4.39) becomes

$$2\left(g'\right)^2 - g g'' = g''' \tag{4.42}$$

$$g(0) = 0, \quad g'(0) = 1, \quad g'(\infty) = 0 \tag{4.43}$$

The solution is $g''(0) = -1.2818085$.

4.2.1.4 The Wall Jet

Consider a momentum jet tangent to (and just above) a wall. The fluid at infinity is quiescent. Unlike a free jet, the momentum is not conserved. Equation (4.16) holds, with $C = 0$ and the boundary conditions are

$$f(0) = 0, \quad f'(0) = 0, \quad f(\infty) = K \tag{4.44}$$

Glauert (1956) integrated Eq. (4.16) from η to infinity to obtain

$$(3m - 1)h + m f f' + v f'' = 0, \quad h = \int_{\eta}^{\infty} \left(f'\right)^2 d\eta \tag{4.45}$$

Now multiply Eq. (4.45) by f' and integrate from zero to infinity, noting $f'(\infty) = 0$.

$$(3m - 1) \int_0^\infty f'h d\eta + m \int_0^\infty f(f')^2 d\eta = 0 \tag{4.46}$$

The second integral in Eq. (4.46) is integrated by parts since $h' = -(f')^2$.

$$\int_0^\infty f(f')^2 d\eta = -fh|_0^\infty + \int_0^\infty hf' d\eta \tag{4.47}$$

But $h(\infty) = 0$. Thus Eq. (4.46) becomes

$$(4m - 1) \int_0^\infty f'h d\eta = 0 \tag{4.48}$$

By assuming f' (the tangential velocity) is non-negative, and since h is positive, Glauert proved

$$m = \frac{1}{4} \tag{4.49}$$

Using the transform $f = Kg(\zeta)$, $\zeta = \frac{K}{\nu}\eta$ Eq. (4.16) yields

$$4g''' + 2(g')^2 + gg'' = 0 \tag{4.50}$$

$$g(0) = 0, \quad g'(0) = 0, \quad g(\infty) = 1 \tag{4.51}$$

The solution is $g''(0) = 0.0138887$ and the universal profile is shown in Fig. 4.3.

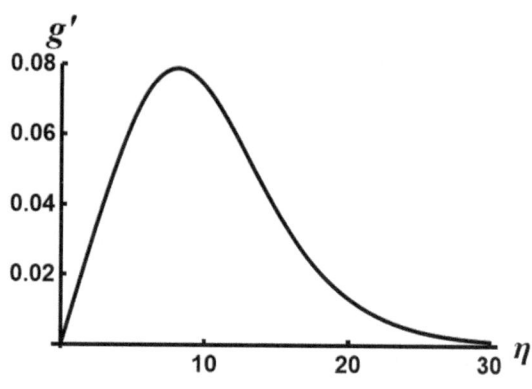

Fig. 4.3 The velocity profile for the wall jet

4.2.2 Axisymmetric Solutions

Most two-dimensional boundary layers have a physical axisymmetric counterpart. Due to the continuity equation, the solution can be quite different. We shall not go into the axisymmetric free jet or the rotating plate, since one can obtain the boundary layer solutions by taking limits on the existing exact N-S solutions.

4.2.2.1 Thin Axisymmetric Boundary Layer

Establish intrinsic coordinates (x, y) along an axisymmetric body as shown in Fig. 4.4a. The boundary layer thickness is assumed to be small compared with the local radius $r_0(x)$. If the surface curvature is also small, the boundary layer momentum conservation equation is the same as the two-dimensional case, i.e.

$$uu_x + vu_y = UU_x + vu_{yy} \tag{4.52}$$

Here (u, v) are velocities in the (x, y) directions respectively. The continuity equation is different. Figure 4.4b shows an elemental ring where, using mass balance similar to Sect. 1.1, we find

$$\frac{\partial}{\partial x}(u\Delta y 2\pi r_0)\Delta x + \frac{\partial}{\partial y}(v\Delta x 2\pi r_0)\Delta y = 0 \tag{4.53}$$

or

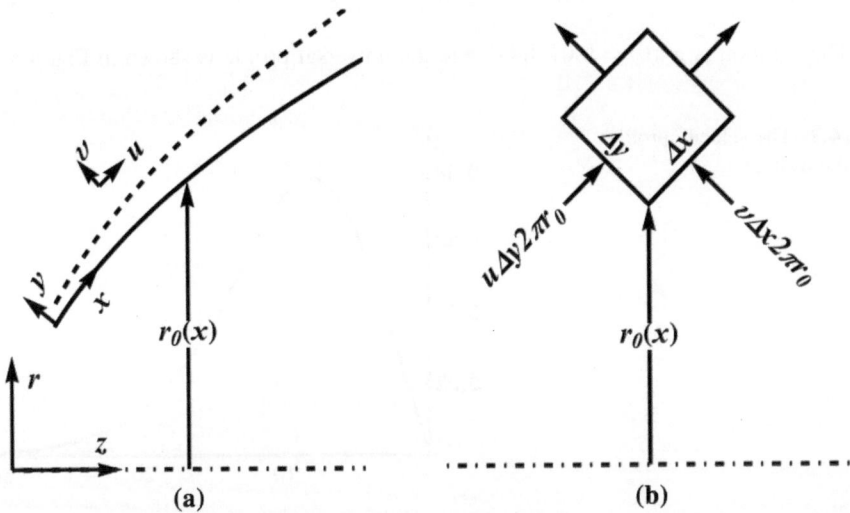

(a)   (b)

Fig. 4.4 a The axisymmetric boundary layer. Dash-dotted line is the symmetry axis **b** An elemental ring showing the mass input

$$\frac{\partial}{\partial x}(ur_0) + \frac{\partial}{\partial y}(vr_0) = 0 \tag{4.54}$$

Define a stream function

$$u = \frac{1}{r_0}\frac{\partial \psi}{\partial y}, \quad v = -\frac{1}{r_0}\frac{\partial \psi}{\partial x} \tag{4.55}$$

Differentiate Eq. (4.52) with respect to y and use Eq. (4.55). The result is

$$\left(\frac{2}{r_0}\right)_x \psi_y \psi_{yy} + \frac{1}{r_0}\frac{\partial(\psi_{yy}, \psi)}{\partial(x, y)} = v\psi_{yyyy} \tag{4.56}$$

Consider the family of power law axisymmetric bodies

$$r_0 = ax^s \tag{4.57}$$

where $a > 0$, $s > 0$. The surface is described parametrically by Eq. (4.57) and

$$z = \int_0^x \sqrt{1 - \left(\frac{dr_0}{dx}\right)^2}\, dx \tag{4.58}$$

Figure 4.5 shows the shapes of the family of axisymmetric bodies with pointed vertices. The body is a cone for $s = 1$ and the opening angles can be adjusted by a.

For a similarity solution let

$$\psi = ax^m f(\eta), \quad \eta = \frac{y}{x^n} \tag{4.59}$$

Equation (4.56) reduces to the similarity equation

$$-2sf'f'' + (m - 2n)f'f'' - mff''' = vf'''' \tag{4.60}$$

provided $m + n = 1 + s$. Integrating Eq. (4.60) once yields

$$(2m - 2s - 1)(f')^2 - mff'' = vf''' + C \tag{4.61}$$

If we let

$$f(\eta) = \bar{f}(\bar{\eta}), \quad m = (1 + 2s)\bar{m}, \quad \bar{\eta} = \frac{\eta}{(1 + 2s)} \tag{4.62}$$

Equation (4.61) becomes

$$(2\bar{m} - 1)(\bar{f}')^2 - \bar{m}\bar{f}\,\bar{f}'' = v\bar{f}''' + \bar{C} \tag{4.63}$$

Fig. 4.5 Various shapes of
power law axisymmetric
bodies, $a = 0.7$. The
dash-dotted line is the axis of
symmetry. From left $s = 1$,
1.4, 2, 3

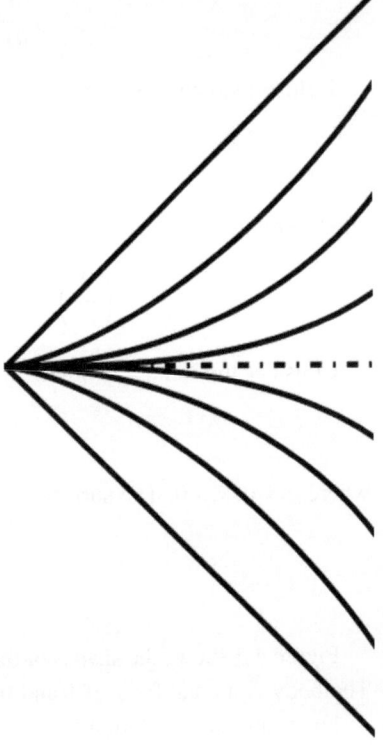

This is the same as the two-dimensional Eq. (4.16). Mangler (1948) devised an integral transform and established the analogy between axisymmetric and two-dimensional boundary layers.

4.2.2.2 The Flow Past a Needle

Consider the axial uniform flow towards a needle, which is defined as a very slender axisymmetric solid. The boundary layer thickness is comparable to the radius of the needle and the governing equation is Eq. (4.11).

By substituting Eq. (A.1) into Eq. (4.11), one can show that similarity solutions are possible when $m = 1$ and arbitrary n. Furthermore, if the outer axial velocity is to be uniform, $n = \frac{1}{2}$. Let

$$\psi = zf(\eta), \quad \eta = \frac{r^2}{z} \tag{4.64}$$

Notice, η is defined differently to simplify the resulting similarity equation. Equation (4.11) gives

$$-\left(f'f'' + ff'''\right) = 2\nu\left(2f''' + \eta f''''\right) \tag{4.65}$$

Integrating once, we have

$$-ff'' = 2v(\eta f''' + f'') \tag{4.66}$$

where we have eliminated the integration constant due to the zero pressure gradient. The boundary conditions are applied on the needle, say, at $\eta = a$, which are paraboloids described by

$$r_0 = \sqrt{az} \tag{4.67}$$

The axial velocity is

$$u = \frac{1}{r}\psi_r = 2f' \tag{4.68}$$

Thus, the boundary conditions are

$$f(a) = 0, \quad f'(a) = 0, \quad f'(\infty) = U/2 \tag{4.69}$$

Using the transform

$$f = vg(\zeta), \quad \zeta = \frac{U}{2v}\eta \tag{4.70}$$

the universal equation is

$$2\zeta g''' + 2g'' + gg'' = 0 \tag{4.71}$$

$$g(b) = 0, \quad g'(b) = 0, \quad g'(\infty) = 1 \tag{4.72}$$

Unfortunately, the nondimensional $b = aU/2v$ cannot be eliminated. Table 4.3 shows the solution for each b. Figure 4.6 shows the axial velocity profile for each needle size. Notice, we need the boundary layer thickness (roughly about 95% of the free stream velocity) to be lager than the radius of the needle ($\zeta = b$ when $g' = 0$). This is indeed true from Fig. (4.6).

4.2.2.3 Flow Caused by an Axisymmetric Boundary Velocity

The flow may be caused by axisymmetric stretching of a bounding surface, or may be caused by spreading of floating material on a fluid surface (Wang 2012). Figure 4.7 shows an axisymmetric layer of material is supplied to the origin with mass rate Q. The layer thickness $h(r)$ may vary due to gravity or surface tension. Mass conservation gives the lateral velocity

$$U(r) = \frac{Q}{2\pi rh} \tag{4.73}$$

Table 4.3 The initial values
for the needle boundary layer

b	$g''(b)$
0.05	5.1547
0.1	3.0063
0.2	1.7815
0.3	1.3223
0.5	0.9165
1	0.5665
2	0.3471
3	0.2750
5	0.1997
10	0.1313

Fig. 4.6 The axial velocity.
From left: $b = 0.05, 1, 2, 3$

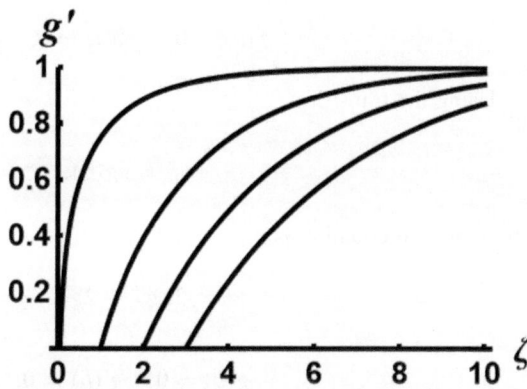

This velocity is imparted to the fluid surface below, causing a boundary layer at the interface.

Equations (4.11, 4.12) give

$$-\frac{1}{r}\frac{\partial(\psi_{zz},\psi)}{\partial(r,z)} + \frac{2}{r^2}\psi_{zz}\psi_z = \nu\psi_{zzzz} \tag{4.74}$$

Substituting

$$\psi = r^m f(\eta), \quad \eta = \frac{z}{r^n} \tag{4.75}$$

we find $n = 2 - m$ and

$$-(3m-4)f'f'' + mff''' + 2f'f'' = \nu f'''' \tag{4.76}$$

Integrate once to obtain

Fig. 4.7 Flow due to spreading of material on a fluid surface. Dashed curves show the boundary layer

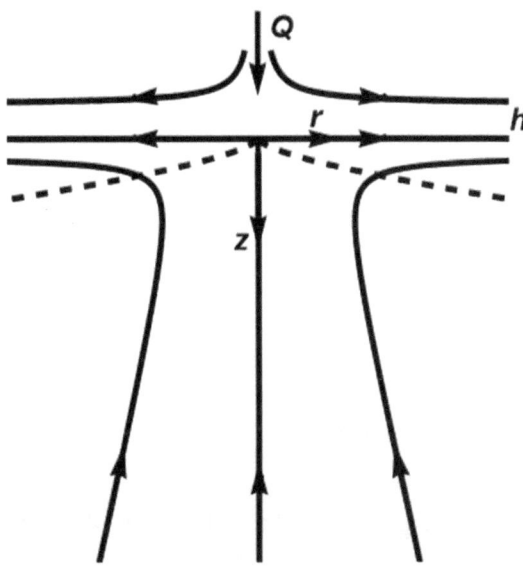

$$(3 - 2m)(f')^2 + mff'' = \nu f''' + C \tag{4.77}$$

The constant C is set to zero since the fluid is quiescent at infinity. The tangential velocity is

$$u = -\frac{1}{r}\psi_z = -r^\alpha f'(\eta), \quad \alpha = 2m - 3 \tag{4.78}$$

If $U = Kr^\alpha$ on the surface, the boundary conditions are

$$f(0) = 0, \quad f'(0) = -K, \quad f'(\infty) = 0 \tag{4.79}$$

The negative sign is due to positive u velocity. The transform

$$f = -\sqrt{\nu K}\,g(\zeta), \quad \zeta = \sqrt{\frac{K}{\nu}}\,\eta \tag{4.80}$$

yields

$$(3 - 2m)(g')^2 + mgg'' + g''' = 0 \tag{4.81}$$

$$g(0) = 0, \quad g'(0) = 1, \quad g'(\infty) = 0 \tag{4.82}$$

Table 4.4 shows the solutions. When $\alpha = -1$, the thickness h is constant. The closed form solution is

Table 4.4 Solutions to axisymmetric spreading on a fluid surface

α	m	$g''(0)$
−1	1	0
0	1.5	−0.76859
1	2	−1.17372
2	2.5	−1.47839

$$g = \sqrt{2}\tanh\left(\zeta/\sqrt{2}\right) \tag{4.83}$$

For $\alpha > -1$, the thickness decreases with the distance from the origin. A physical interpretation for $\alpha = 1$ is the constant, even deposition of rain or condensation on the surface, for which $U \sim r$.

4.3   Approximate Solutions of Boundary Layers

When the boundary layer equation does not have an exact or similarity solution, approximate methods may be considered. These include integral methods which give global properties but may not be detailed enough for the flow field or the determination of transport properties. Another approximation involves series expansion along the boundary layer, such as the analysis of the boundary layer development of the flow inside (or outside) a cylinder. However, these methods are very tedious and invariably require some numerical computation or empirical assumption. In addition, these methods are somewhat outdated since such problems can now be more easily treated by a numerical integration code.

Instead, we shall present here the far wake problem, which is an approximation of the boundary layer equation (itself being an approximation).

4.3.1   Two-Dimensional Wake

A wake is produced behind a body in a uniform stream with velocity U. Balancing force, the drag of the body is equal to the momentum deficiency across a transverse plane in the wake.

$$D = \rho \int_{-\infty}^{\infty} u\tilde{u}\,dy \tag{4.84}$$

Here, u is the longitudinal velocity and $\tilde{u}$ is the velocity deficiency.

$$\tilde{u} = U - u \tag{4.85}$$

In terms of stream function, Eq. (4.85) is

$$\psi = Uy - \tilde{\psi} \tag{4.86}$$

Assume $\tilde{\psi} \ll Uy$, which is true in the far wake. Equation (4.1) then linearizes to

$$U\tilde{\psi}_{yyx} = \nu\tilde{\psi}_{yyyy} \tag{4.87}$$

Noting that the perturbation velocity $\tilde{\psi}_y$ is zero at infinity, Eq. (4.87) is integrated twice to obtain

$$U\tilde{\psi}_x = \nu\tilde{\psi}_{yy} \tag{4.88}$$

Equation (4.84) linearizes to

$$D = \rho \int_{-\infty}^{\infty} U\tilde{\psi}_y dy = 2\rho U\tilde{\psi}(\infty) \tag{4.89}$$

where we have used the symmetry property in y. Substituting Eq. (4.13) into Eqs. (4.88, 4.89) we find $m = 0$, $n = \frac{1}{2}$ and

$$\nu f''(\eta) + \frac{1}{2}U\eta f'(\eta) = 0, \quad f(0) = 0, \quad f'(\infty) = 0 \tag{4.90}$$

The solution is

$$\tilde{\psi} = f = c\sqrt{\frac{\nu\pi}{U}}\,\mathrm{erf}\left(\sqrt{\frac{U}{4\nu}}\eta\right) \tag{4.91}$$

Here, erf is the error function and using Eq. (4.89), the velocity is

$$\tilde{u} = \frac{c}{\sqrt{x}}e^{-\frac{U}{4\nu}\eta^2}, \quad c = \frac{D}{2\rho\sqrt{\nu\pi U}}, \quad \eta = \frac{y}{\sqrt{x}} \tag{4.92}$$

4.4 Matched Asymptotic Expansions

We illustrate higher order corrections of the boundary layer solution using matched asymptotic expansions (Appendix B). Since several different coordinate systems are used, it is easier to construct the solution step by step instead of a formal asymptotic expansion.

Fig. 4.8 The scraping problem. Boundary layers are indicated by dashed curves

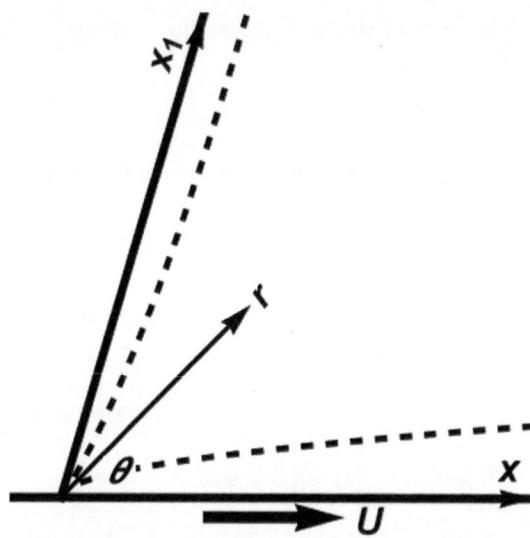

4.4.1 The Scraping Problem

Figure 4.8 shows the scraping problem. Placing cylindrical coordinates (r, θ) at the point of contact, the horizonal plate at $\theta = 0$ moves with velocity U, and the scraper is the slanted fixed plate at $\theta = \beta$. The primary boundary layer on the moving plate is also the Sakiadis layer of Sect. 4.2.1.3.

From Sect. 4.2.1.3, the primary (Sakiadis) layer is

$$\psi = \sqrt{x} f(\eta), \quad \eta = \frac{y}{\sqrt{x}}, \quad f = \sqrt{\nu U} g(\zeta), \quad \zeta = \sqrt{\frac{U}{\nu}} \eta \qquad (4.93)$$

Then,

$$2g''' + gg'' = 0, \quad g(0) = 0, \quad g'(0) = 1, \quad g'(\infty) = 0 \qquad (4.94)$$

The solution is

$$g''(0) = -0.443748, \quad g(\infty) = 1.61612 \equiv c \qquad (4.95)$$

Therefore, the boundary layer produces a displacement $g(\infty)$, which forces a higher order outer flow. Let $\Psi(r, \theta)$ be the outer stream function. Preliminary matching gives

$$\Psi(r, 0) = \psi(x, \infty) = c\sqrt{\nu U} \sqrt{x} \qquad (4.96)$$

The potential solution, which is also zero on the fixed plate at $\theta = \beta$, is

$$\Psi = c\sqrt{\nu U}\sqrt{r}\frac{\sin\left(\frac{\beta-\theta}{2}\right)}{\sin\left(\frac{\beta}{2}\right)} \tag{4.97}$$

But this outer flow cannot satisfy the no-slip condition on the fixed plate. Let subscript 1 denote the secondary boundary layer on the fixed plate. The imparted tangential velocity is

$$U_1 = \frac{1}{r}\Psi_\theta(r,\beta) = \frac{-c\sqrt{\nu U}}{\sin\left(\frac{\beta}{2}\right)}r^{-1/2} < 0 \tag{4.98}$$

The inner velocity satisfies Eq. (4.1). The boundary condition from matching Eq. (4.98) is

$$\psi_{1y_1} \sim -c_1 x_1^{-1/2}, \quad c_1 \equiv \frac{c\sqrt{\nu U}}{\sin\left(\frac{\beta}{2}\right)} > 0 \tag{4.99}$$

Using Eqs. (4.13, 4.14) and Eq. (4.99), we find $m = \frac{1}{4}$, $n = \frac{3}{4}$ and

$$-\frac{5}{4}f_1 f_1'' - \frac{1}{4}f_1 f_1''' = \nu f_1'''' \tag{4.100}$$

Integrate once and use the transform

$$f_1(\eta_1) = -\sqrt{\nu c_1}g_1(\zeta_1), \quad \zeta_1 = \sqrt{\frac{c_1}{\nu}}\eta_1 \tag{4.101}$$

The universal equation, after integrating once, is

$$2[(g_1')^2 - 1] + g_1 g_1'' = 4g_1''', \quad g_1(0) = 0, \quad g_1'(0) = 0, \quad g_1'(\infty) = 1 \tag{4.102}$$

The solution is $g_1''(0) = 0.791152555$. We shall not go into the higher order corrections which are due to the displacement of the secondary boundary layer, which in turn leads to the secondary outer flow and further corrections to the primary boundary layer.

4.5 Some Comments

The analytic methods presented apply to laminar boundary layers. The results may be applied to turbulent boundary layers, if the viscosity is replaced by an effective turbulent viscosity and adjusted for its possible variation.

For each given outer tangential velocity a boundary layer may be constructed to satisfy the no-slip condition. In order for the solution to be physically significant, the outer velocity must be from a plausible potential flow.

Fig. 4.9 Two interacting shear
flows behind a splitter plate

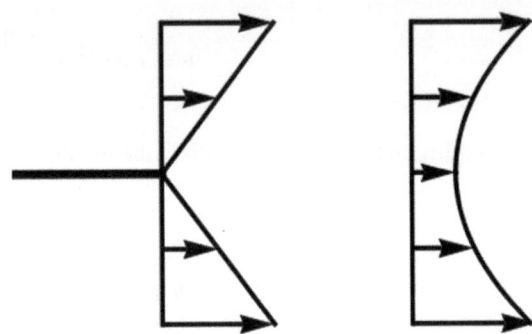

Boundary layers may separate from the solid surface when encountering adverse pressure gradients. Separation also causes large wakes, which significantly alter the original assumed outer potential flow. Thus, it may be improper to apply boundary layer methods for the flow over blunt bodies such as cylinders or spheres. For these cases, numerical (CFD) codes should be used.

However, boundary layers can be thin for blunt bodies in some unsteady flow situations, as we shall see in Chap. 5.

Exercises

(4.1) Find the stress on the plate in Blasius flow (use Chap. 1 formulas). What would be the drag on a finite plate of length L? What is the minimum velocity at a distance $3L$ form the trailing edge?

(4.2) Suppose two symmetric linear shear flows are brought together (Fig. 4.9). Formulate the universal equation and solve it by a desk computer.

(4.3) Set up and solve the problem of an axisymmetric momentum boundary layer which emanates from a hole. Can you compare your solution to the Landau jet of Sect. 2.2.6?

(4.4) Solve the axisymmetric wake problem similar to Sect. 4.3.1.

(4.5) Try using successive matched asymptotic expansions for the higher order corrections of the Blasius boundary layer.

Notes

The best overall references for boundary layers are Rosenhead (1963), Schlichting and Gersten (2000) and White (2006). A good overview of stretching boundary layers was by Banks (1983). References for two-dimensional similarity boundary layers are Falkner and Skan (1931), Blasius (1908), Lock (1951), Schlichting (1933), Wang (1992), Sakiadis (1961) and Glauert (1956). Exponential similarity boundary layer was discussed by Jones and Watson (1963) and Magyari and Keller (1999). The Mangler transform is most accessible at Mangler (1948). The needle boundary layer was considered by Lee (1967). The boundary layer due to axisymmetric spreading of material was discussed in Wang (2012). The wake problem was discussed in Berger (1971) and Wang (1989).

References

H.H. Banks, J. Mech. Theor. Appl. **2**, 375–392 (1983)

S.A. Berger, *Laminar Wakes* (Elsevier, NY, 1971)

H. Blasius, ZAMP **56**, 1–37 (1908)

V.M. Falkner and S.W. Skan, Phil. Mag. **12**, 865–896 (1931)

M.B. Glauert, J. Fluid Mech. **1**, 625–643 (1956)

C.W. Jones and E.J. Watson, in *Laminar Boundary Layers* L. Rosenhead, Ed. Clarendon, Oxford, 1963.

L.L. Lee, Phys. Fluids **10**, 820–822 (1967)

R.C. Lock, Quart. J. Mech. Appl. Math. **4**, 42–63 (1951)

E. Magyari, B. Keller, J. Phys. D. Appl. Phys. **32**, 577–585 (1999)

W. Mangler, ZAMM **28**, 97–103 (1948)

L. Rosenhead (ed.), *Laminar Boundary Layers* (Clarendon, Oxford, 1963)

B.C. Sakiadis, AIChE J. **7**, 221–225 (1961)

H. Schlichting, ZAMM **13**, 260–263 (1933)

H. Schlichting, K. Gersten, *Boundary Layer Theory*, 8th edn. (Springer, Berlin, 2000)

C.Y. Wang, Mech. Res. Comm. **16**, 219–225 (1989)

C.Y. Wang, Phys. Fluids A **4**, 1304–1306 (1992)

C.Y. Wang, J. Math. Anal. Appl. **385**, 1190–1194 (2012)

F.M. White, *Viscous Fluid Flow*, 3rd edn. (McGraw-Hill, Boston, 2006)

In this chapter we consider problems where unsteady effects are dominant. The phenomena are quite different from steady or quasi-steady viscous flows.

We shall not consider exact unsteady solutions of the N-S equation which were reviewed in the references of Chap. 2. We shall also exclude unsteady parallel flows, which are analogous to the well-known unsteady diffusion of heat. In what follows, we shall present several representative examples of unsteady viscous flows.

5.1 Impulsive Start of a Circular Cylinder

For the edgewise impulsive start of a semi-infinite plate, the section close to the leading edge is like an unsteady Blasius boundary layer while further downstream, it is Stokes' first (Rayleigh) problem for an infinite plate. As in the entrance flow problem, it is difficult to connect these two regions.

But analytic solutions for the initial phase of starting of blunt bodies may be possible. Let us illustrate with the impulsive start of a circular cylinder at high but laminar Reynolds numbers. Previous literature ignored the curvature and the effects of outer flow, both important in the first order correction. We shall properly use cylindrical coordinates and matched asymptotic expansions (Appendix B).

Experiments show the flow is potential at the beginning with a very thin boundary layer, since vorticity needs time to diffuse. Reverse flow then occurs within the attached boundary layer and finally vorticity becomes detached, and a large wake is formed at steady state.

Consider a circular cylinder suddenly starts to move with a constant transverse velocity U. If the coordinates are fixed on the body, then the flow moves with relative velocity $-U$. Let L be the radius of the cylinder. A normalization of Eq. (1.36), similar to Eq. (3.5), gives

$$S \frac{\partial}{\partial t} \nabla^2 \psi + \frac{1}{r} \frac{\partial (\nabla^2 \psi, \psi)}{\partial (r, \theta)} = \frac{1}{R} \nabla^4 \psi \qquad (5.1)$$

where

$$S = \frac{L}{UT}, \quad R = \frac{UL}{\nu} \qquad (5.2)$$

Assume both S and R are large, so that

$$\varepsilon = \frac{1}{S} \ll 1, \quad \frac{1}{R} = \alpha \varepsilon \ll 1 \qquad (5.3)$$

Here, $\alpha = O(1)$ in order to bring up the importance of both unsteady and viscous effects. The physical meaning of large S is that the time scale T considered is small compared to the time required to travel one radius. Thus, the N-S equation is

$$\frac{\partial}{\partial t} \nabla^2 \psi + \frac{\varepsilon}{r} \frac{\partial (\nabla^2 \psi, \psi)}{\partial (r, \theta)} = \alpha \varepsilon^2 \nabla^4 \psi \qquad (5.4)$$

For the outer flow, let

$$\psi = \Psi_0 + \varepsilon \Psi_1 + \cdots \qquad (5.5)$$

We find the outer flow is potential and independent of time for two orders

$$\nabla^2 \Psi_0 = 0, \quad \nabla^2 \Psi_1 = 0 \qquad (5.6)$$

The zeroth order solution with uniform velocity $-U$ at infinity and no penetration on the cylinder at $r = 1$ is

$$\Psi_0 = -\sin(\theta) \left(r - \frac{1}{r} \right) \qquad (5.7)$$

In the boundary layer let

$$r = 1 + \varepsilon \hat{r}, \quad \hat{r} = O(1) \qquad (5.8)$$

Equation (5.7) becomes

$$\Psi_0 = -\sin(\theta) (2\varepsilon \hat{r} + \cdots) \qquad (5.9)$$

Since $\Psi_0 = O(\varepsilon)$ in the boundary layer, the inner expansion is

$$\psi = \varepsilon \psi_0 + \varepsilon^2 \psi_1 + \cdots \qquad (5.10)$$

Substituting into Eq. (5.4) and noting

$$\nabla^2 = \frac{1}{\varepsilon^2}\frac{\partial^2}{\partial \hat{r}^2} + \frac{1}{\varepsilon}\frac{\partial}{\partial \hat{r}} + \cdots \tag{5.11}$$

we obtain the leading order

$$\psi_{0\hat{r}\hat{r}t} = \alpha \psi_{0\hat{r}\hat{r}\hat{r}\hat{r}} \tag{5.12}$$

The solution of Eq. (5.12) is the combination of the solutions of

$$\psi_{0t} = \alpha \psi_{0\hat{r}\hat{r}} \tag{5.13}$$

and

$$\psi_{0\hat{r}\hat{r}} = 0 \tag{5.14}$$

Now Eq. (5.13) is a diffusion equation similar to Stokes' first problem. The similarity variable is

$$\eta = \frac{\hat{r}}{2\sqrt{\alpha t}} \tag{5.15}$$

Equation (5.9) yields

$$\Psi_0 = -\sin(\theta)(2\varepsilon\hat{r} + \cdots) = -c_0\varepsilon\sqrt{t}\eta + \cdots, \quad c_0 = 4\sqrt{\alpha}\sin(\theta) \tag{5.16}$$

To match the outer solution, the inner stream function must be of the form

$$\psi_0 = -c_0\sqrt{t}f(\eta) \tag{5.17}$$

Substitution of Eq. (5.17) into Eq. (5.13) yields

$$f'' - 2(f - \eta f') = 0 \tag{5.18}$$

The solution, with those from Eq. (5.14), and together satisfy the no-slip boundary conditions, is

$$f = \eta - \eta\,\mathrm{erfc}(\eta) - \frac{1}{\sqrt{\pi}}\left(1 - e^{-\eta^2}\right) \tag{5.19}$$

where erfc is the complementary error function. Matching Eq. (5.10) with Eq. (5.5) results in

$$\Psi_1|_{r=1} = c_0\sqrt{t}\frac{1}{\sqrt{\pi}} = 4\sqrt{\frac{\alpha t}{\pi}}\sin(\theta) \tag{5.20}$$

Fig. 5.1 Instantaneous
streamlines showing rear
recirculating eddies.
Dash-dotted line is the
symmetry plane

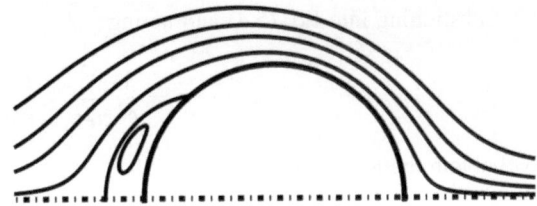

The first order correction to the outer flow is the doublet, which causes a radial displacement

$$\Psi_1 = 4\sqrt{\frac{\alpha t}{\pi}} \frac{\sin(\theta)}{r} \tag{5.21}$$

This flow induces its own correction to the boundary layer near the cylinder. After some work, the composite solution is

$$
\psi = -\sin(\theta)\left(r - \frac{1}{r}\right) + \varepsilon 4\sqrt{\frac{\alpha t}{\pi}} \frac{\sin(\theta)}{r} + \varepsilon 4\sqrt{\alpha t}\,\sin(\theta)\left[\eta\,\mathrm{erfc}(\eta) - \frac{1}{\sqrt{\pi}}e^{-\eta^2}\right]
$$
$$
- \varepsilon^2 4 \frac{\alpha t}{\sqrt{\pi}}\sin(\theta)\left\{\frac{\sqrt{\pi}}{4}[1 - \mathrm{erfc}(\eta)] - \frac{3\sqrt{\pi}}{2}\eta^2\mathrm{erfc}(\eta) + \frac{3}{2}\eta e^{-\eta^2}\right\}
$$
$$
+ \varepsilon^2 8\sqrt{\alpha}t^{3/2}\sin(\theta)\cos(\theta)F(\eta) + O\left(\varepsilon^3\right) \tag{5.22}
$$

The function $F(\eta)$ is complicated, but given in e.g. Wang (1967). Figure 5.1 shows the streamlines at a short time after impulsive start. Notice the rear attached eddy.

5.2   Acoustic Streaming

A sound field whose wavelength is much larger than an imbedded solid body is equivalent to an incompressible oscillating flow over such a body. Due to the nonlinearity of the N-S equation, a purely oscillatory flow may generate a steady flow called acoustic streaming. This steady flow can be utilized for mixing or transport in micro-fluidics.

The basic theory for acoustic streaming is as follows. Equation (3.5) is

$$S\nabla^2\psi_t + \frac{\partial(\nabla^2\psi, \psi)}{\partial(x, y)} = \frac{1}{R}\nabla^4\psi \tag{5.23}$$

Let the stream function be the sum of an unsteady part denoted by a tilde and a steady part denoted by an overbar.

$$\psi = \tilde{\psi} + \overline{\psi} \tag{5.24}$$

Equation (5.23) can then be separated into unsteady and steady equations

$$SV^2\tilde{\psi}_t + \frac{\partial\left(\nabla^2\tilde{\psi},\overline{\psi}\right)}{\partial(x,y)} + \frac{\partial\left(\nabla^2\overline{\psi},\tilde{\psi}\right)}{\partial(x,y)} + \frac{\widetilde{\partial\left(\nabla^2\tilde{\psi},\tilde{\psi}\right)}}{\partial(x,y)} = \frac{1}{R}\nabla^4\tilde{\psi} \tag{5.25}$$

$$\frac{\partial\left(\nabla^2\overline{\psi},\overline{\psi}\right)}{\partial(x,y)} + \frac{\overline{\partial\left(\nabla^2\tilde{\psi},\tilde{\psi}\right)}}{\partial(x,y)} = \frac{1}{R}\nabla^4\overline{\psi} \tag{5.26}$$

The tilde and overbar of a product of periodic functions are shown in the example

$$\cos(t)\cos(t) = \frac{1}{2} + \frac{1}{2}\cos(2t) \tag{5.27}$$

Thus,

$$\widetilde{\cos(t)\cos(t)} = \frac{1}{2}\cos(2t), \quad \overline{\cos(t)\cos(t)} = \frac{1}{2} \tag{5.28}$$

The steady streaming is generated by the second term in Eq. (5.26), which is the steady part of the product of two oscillatory functions.

5.2.1 Streaming at a Stagnation Point

Steady streaming occurs when a body (e.g. a cylinder) is placed in an oscillating fluid. Since an oscillatory flow parallel to a surface does not induce any steady flow (Stokes second problem, Sect. 2.1.2), the steady streaming must be caused by the oscillatory flow normal to a surface near the stagnation point (Wang 1968).

Consider oscillatory stagnation flow towards an infinite plane (Fig. 5.2a). Similar to the steady Hiemenz exact solution (Sect. 2.2.2), far from the plate, the normal velocity is prescribed as $-ay^*\cos(\omega t^*)$, where a is an amplitude of dimension $[1/T]$ and ω is the oscillation frequency also of $[1/T]$. There is no natural length scale, thus $L = \sqrt{v/a}$. Assuming high frequency and low Reynolds numbers, the two non-dimensional parameters are

$$S = \frac{\omega}{a} \equiv \frac{1}{\varepsilon^2} \gg 1, \quad R = 1 \tag{5.29}$$

However, we do not know the order of the steady flow. Let

$$O(\overline{\psi}) = \gamma O(\tilde{\psi}) \tag{5.30}$$

where the order of $\gamma < O(1)$ is to be determined. Equations (5.25, 5.26) become

$$\frac{1}{\varepsilon^2}\nabla^2\tilde{\psi}_t + \gamma\frac{\partial\left(\nabla^2\tilde{\psi},\overline{\psi}\right)}{\partial(x,y)} + \gamma\frac{\partial\left(\nabla^2\overline{\psi},\tilde{\psi}\right)}{\partial(x,y)} + \frac{\widetilde{\partial\left(\nabla^2\tilde{\psi},\tilde{\psi}\right)}}{\partial(x,y)} = \nabla^4\tilde{\psi} \tag{5.31}$$

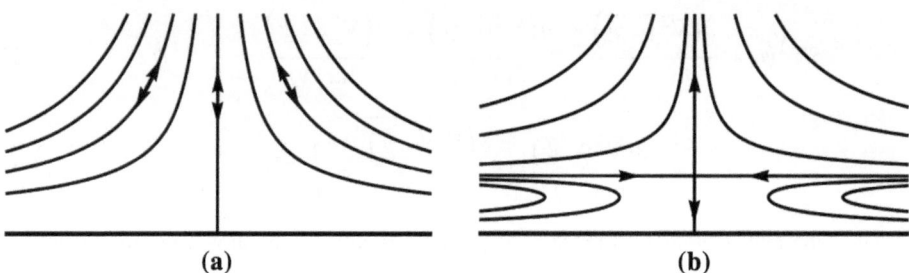

Fig. 5.2 a Oscillatory stagnation flow towards a plate **b** Generated steady streaming

$$\gamma^2 \frac{\partial(\nabla^2 \overline{\psi}, \overline{\psi})}{\partial(x, y)} + \overline{\frac{\partial(\nabla^2 \tilde{\psi}, \tilde{\psi})}{\partial(x, y)}} = \gamma \nabla^4 \overline{\psi} \tag{5.32}$$

From Eq. (5.31), the unsteady flow has a boundary layer of order ε near the plate, and it is potential outside the layer. Thus, the forcing term in Eq. (5.32) only exists inside the boundary layer where

$$y = \varepsilon \eta \tag{5.33}$$

Equation (5.26) shows $\gamma = \varepsilon^2$.

Now expand the outer unsteady stream function

$$\tilde{\psi} = \tilde{\psi}_0 + \varepsilon \tilde{\psi}_1 + \cdots \tag{5.34}$$

Equation (5.31) and the conditions at infinity show $\tilde{\psi}_0$ is potential

$$\tilde{\psi}_0 = -xye^{it} \tag{5.35}$$

where $i = \sqrt{-1}$ and only the real part has any physical significance. In the boundary layer

$$\tilde{\psi}_0 \sim -x\varepsilon\eta e^{it} \tag{5.36}$$

Thus, the inner unsteady stream function expansion is

$$\tilde{\psi} = \varepsilon \tilde{\psi}_0 + \varepsilon^2 \tilde{\psi}_1 + \cdots \tag{5.37}$$

Equation (5.31) yields

$$\tilde{\psi}_{0\eta\eta t} = \tilde{\psi}_{0\eta\eta\eta\eta} \tag{5.38}$$

The solution that matches Eq. (5.36) and satisfies the no-slip conditions on the plate is

$$\tilde{\psi}_0 = -x e^{it}\left[\eta - \frac{(1-i)}{\sqrt{2}}\left(1 - e^{-\zeta}\right)\right], \quad \zeta = \frac{(1+i)}{\sqrt{2}}\eta \tag{5.39}$$

Matching with the outer solution again gives

$$\tilde{\psi}_1 = \frac{(1-i)}{\sqrt{2}} x y e^{it} \tag{5.40}$$

For the steady equation, let

$$\overline{\psi} = \varepsilon\overline{\psi}_0 + \varepsilon^2\overline{\psi}_1 + \cdots \tag{5.41}$$

The leading order of Eq. (5.32) is

$$\frac{1}{2}\left(\tilde{\psi}_{0\eta\eta x}^* \tilde{\psi}_{0\eta} - \tilde{\psi}_{0\eta\eta\eta}^* \tilde{\psi}_{0x}\right) = \overline{\psi}_{0\eta\eta\eta\eta} \tag{5.42}$$

Here the asterisk denotes the complex conjugate. The solution that satisfies the no-slip condition is

$$\overline{\psi}_0 = -x\left\{\frac{3}{4}\eta - \frac{13}{4\sqrt{2}} + \frac{1}{4\sqrt{2}}e^{-\sqrt{2}\eta} + \left[\frac{3}{\sqrt{2}}\cos\left(\frac{\eta}{\sqrt{2}}\right) + \left(\sqrt{2} + \frac{\eta}{2}\right)\sin\left(\frac{\eta}{\sqrt{2}}\right)\right]e^{-\frac{\eta}{\sqrt{2}}}\right\} \tag{5.43}$$

This steady boundary layer solution generates an outer steady flow

$$\overline{\Psi}_0 = -\frac{3}{4}xy \tag{5.44}$$

The composite steady solution is shown in Fig. 5.2b. Aside from the cells close to the plate, the steady streaming is away from the stagnation point. If a cylinder is placed in an oscillating stream, the generated external steady streaming will be moving away from the two stagnation points.

5.2.2 Streaming Due to an Oscillatory Stretching Boundary

Consider an elastic sheet which stretches back and forth in an otherwise quiescent fluid (Wang 1988). Would steady streaming be generated? Let the tangential velocity of the sheet be $ax^*\cos(\omega t^*)$. An analysis similar to that of the previous section shows the unsteady flow is mostly confined in a boundary layer. Equation (5.39) is replaced by

$$\tilde{\psi}_0 = x e^{it}\frac{(1-i)}{\sqrt{2}}\left(1 - e^{-\zeta}\right) \tag{5.45}$$

Solving Eq. (5.42) gives

Fig. 5.3 Steady streaming
caused by an oscillatory
stretching sheet

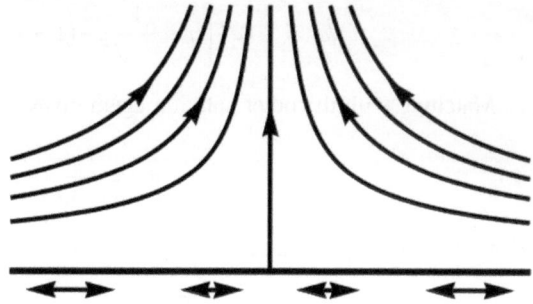

$$\overline{\psi}_0 = x\left\{-\frac{3}{4}\eta + \frac{3}{4\sqrt{2}} - \frac{1}{4\sqrt{2}}e^{-\sqrt{2}\eta} - \frac{1}{2\sqrt{2}}\left[\cos\left(\frac{\eta}{\sqrt{2}}\right) - \sin\left(\frac{\eta}{\sqrt{2}}\right)\right]e^{-\frac{\eta}{\sqrt{2}}}\right\} \quad (5.46)$$

The composite steady streamlines are shown in Fig. 5.3. Notice recirculating cells are absent.

5.3   Squeeze Problems

The squeezing of a viscous fluid from two plates in relative motion models skidding and braking problems. The squeezing from a tube occurs in biological vasoconstriction. We shall illustrate by the two-dimensional braking problem.

Figure 5.4 shows the top plate is squeezing down by the force F and the bottom plate is moving in its own plane with velocity U^*. The distance between the plates is $aZ(t^*)$, where a is the amplitude and Z, of order one, is a decreasing function of the time t^*.

Normalize all lengths by a, the time by a typical squeeze time T, the velocity by a/T and the stream function by a^2/T. The time scale depends on the problem. For example, consider the constant velocity squeezing with speed V^*. We find $Z = 1 - t$. The two plates touch when $t = 1$ or the time scale is $T = a/V^*$.

Equation (1.21) becomes

$$S\left[\nabla^2\psi_t + \frac{\partial(\nabla^2\psi, \psi)}{\partial(x, y)}\right] = \nabla^4\psi \quad (5.47)$$

Fig. 5.4 The unsteady
squeezing of two plates
enclosing a viscous fluid

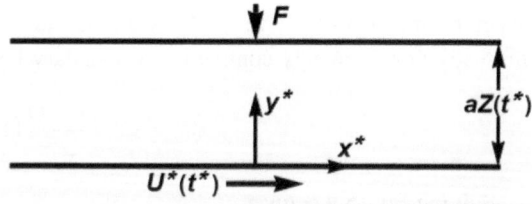

Here, $S = a^2/\nu T$ is the "squeeze number" representing the importance of gap width to viscosity. Our experience in squeeze flows suggests

$$\psi = x Z'(t) f(\eta, t) + Z U(t) g(\eta, t) \tag{5.48}$$

where

$$\eta = \frac{y}{Z(t)}, \quad \frac{\partial F(\eta, t)}{\partial y} = \frac{1}{Z}\frac{\partial F}{\partial \eta}, \quad \frac{\partial F(\eta, t)}{\partial t} = -\frac{Z'}{Z}\eta\frac{\partial F}{\partial \eta} + \frac{\partial F}{\partial t} \tag{5.49}$$

Equation (5.47) can be separated into the terms with x

$$S\left[\left(\frac{Z'}{Z^2}\right)' f_{\eta\eta} + \frac{Z'}{Z^2} f_{\eta\eta t} + \frac{Z'^2}{Z^3}\left(f_{\eta\eta} f_\eta - f_{\eta\eta\eta} f - \eta f_{\eta\eta\eta}\right)\right] = \frac{Z'}{Z^4} f_{\eta\eta\eta\eta} \tag{5.50}$$

and terms without x

$$S\left[\left(\frac{U}{Z}\right)' g_{\eta\eta} + \frac{U}{Z} g_{\eta\eta t} + \frac{U Z'}{Z^2}\left(f_{\eta\eta} g_\eta - g_{\eta\eta\eta} f - \eta g_{\eta\eta\eta}\right)\right] = \frac{U}{Z^3} g_{\eta\eta\eta\eta} \tag{5.51}$$

The boundary conditions on the bottom plate are that the vertical velocity is zero and that the horizontal velocity is uniform $U(t)$

$$f(0, t) = 0, \quad g(0, t) = 0, \quad f_\eta(0, t) = 0, \quad g_\eta(0, t) = 1 \tag{5.52}$$

The boundary conditions on the top plate are that the horizontal velocity is zero and that the vertical velocity is

$$V^* = -\frac{d}{dt^*}\left[a Z(t^*)\right] \tag{5.53}$$

or

$$f(1, t) = 1, \quad g(1, t) = 0, \quad f_\eta(1, t) = 0, \quad g_\eta(1, t) = 0 \tag{5.54}$$

Equations (5.50, 5.51) are unsolvable analytically. But a perturbation solution is possible for small squeeze number S, which implies small gap width. Expand in asymptotic series

$$f(\eta, t) = f_0 + S f_1 + S^2 f_2 + \cdots, \quad g(\eta, t) = g_0 + S g_1 + S^2 g_2 + \cdots \tag{5.55}$$

Then the solution can be obtained for given plate motion $Z(t), U(t)$. Notice that for small S, the time derivative does not enter the zeroth order equations, thus, the initial transient is not important. The zeroth order equations are

$$f_{0\eta\eta\eta\eta} = 0, \quad g_{0\eta\eta\eta\eta} = 0 \tag{5.56}$$

Together with the zeroth order boundary conditions, the solutions are

$$f_0 = 3\eta^2 - 2\eta^3, \quad g_0 = \eta - 2\eta^2 + \eta^3 \tag{5.57}$$

The first order solutions are

$$f_1 = \frac{Z^4}{Z'}\left(\frac{Z'}{Z^2}\right)'\left(\eta^2 - 4\eta^3 + 5\eta^4 - 2\eta^5\right)/20 + ZZ'\left(15\eta^2 - 24\eta^3 + 14\eta^5 - 7\eta^6 + 2\eta^7\right)/35$$

$$g_1 = \frac{Z^3}{U}\left(\frac{U}{Z}\right)'\left(-4\eta^2 + 11\eta^3 - 10\eta^4 + 3\eta^5\right)/60 + ZZ'\left(\begin{array}{c}-69\eta^2 + 67\eta^3 + 105\eta^4 \\ -147\eta^5 + 56\eta^6 - 12\eta^7\end{array}\right)/420$$

$$\tag{5.58}$$

We leave the details of finding the pressure which balances the force on the top plate and the shear stress which affects the motion of the bottom plate to the references. The possible scenarios are (1) constant velocity braking (2) constant force braking and (3) constant power braking. See Wang (1980, 1981).

Exercises

(5.1) Find the solution to an impulsively accelerated cylinder.

(5.2) Find the solution to an impulsively started sphere.

(5.3) Find the steady force on a sphere adjacent to an oscillatory source.

(5.4) Study the steady streaming caused by a large disk torsionally oscillating in its own plane.

(5.5) Solve the problem of squeezing of a fluid between two circular disks.

Notes

A good reference on unsteady viscous flows is Telionis (1981). A good discussion on the impulsively started semi-infinite plate is also in Sherman (1990). Previous literature on impulsively started cylinder or sphere includes Schlichting and Gersten (2000), Wang (1967a, b). A good review of acoustic streaming was given by Wang (1968). The oscillatory stretching of a sheet was by Wang (1988). Squeezing and braking problems were studied by Wang (1980, 1981).

References

F.S. Sherman, *Viscous Flow* (McGraw-Hill, NY, 1990)
H. Schlichting and K. Gersten, *Boundary Layer Theory*, 8th Ed. (Springer, Berlin, 2000)
D.P. Telionis, *Unsteady Viscous Flows* (Springer, NY, 1981)

C.Y. Wang, Acta Mech. **72**, 261–268 (1988)
C.Y. Wang, Int. J. Eng. Sci. **19**, 891–900 (1981)
C.Y. Wang, J. Appl. Mech. **34**, 823–828 (1967a)
C.Y. Wang, J. Math. Phys. **46**, 195–202 (1967b)
C.Y. Wang, J. Fluid Mech. **32**, 55–68 (1968)
C.Y. Wang, ZAMP **31**, 620–627 (1980)

Stokes Flow

6

In Stokes flow or creeping flow, the viscous forces dominate the inertial forces. Since the Reynolds number is small, the basic equation for the stream function is the linear biharmonic equation. We shall exclude parallel or concentric flows for which the inertial terms are identically zero due to geometry.

We shall not discuss complex variable solutions of Stokes flow. Unlike the harmonic equation of potential flow, the biharmonic equation does not have the conformal mapping property. There exists however a circle theorem of introducing a circular cylinder in a flow (and a sphere theorem in three dimensions) under restricted conditions.

We shall illustrate analytic solutions of Stokes flow through some representative examples.

6.1 General Solutions for Steady Stokes Flow

The stream function in steady Stokes flow satisfies (Eq. 3.7)

$$\nabla^4 \psi = 0 \tag{6.1}$$

Since the biharmonic equation is linear, the method of separation of variables can be used.

In two-dimensional Cartesian coordinates, the solution is the combination of

$$\psi = \{\sin(ax), \cos(ax)\}\{e^{ay}, e^{-ay}, ye^{ay}, ye^{-ay}\}, \; a \neq 0$$
$$\{e^{ax}, e^{-ax}\}\{\sin(ay), \cos(ay), y\sin(ay), y\cos(ay)\}, \; a \neq 0$$
$$\{1, x\}\{1, y, y^2, y^3\}, \; a = 0 \tag{6.2}$$

In Eq. (6.2), the variables x and y can be interchanged and a could be complex.

© The Author(s), under exclusive license to Springer Nature Switzerland AG 2024
C. Y. Wang, *Essential Analytic Laminar Flow*, Synthesis Lectures on Engineering, Science, and Technology, https://doi.org/10.1007/978-3-031-36449-5_6

In two-dimensional cylindrical coordinates (r, θ), the solution is the linear combination of

$$\psi = \{r^{1+s}, r^{1-s}\}\{\cos[(1+s)\theta], \sin[(1+s)\theta], \cos[(1-s)\theta], \sin[(1-s)\theta]\}, \quad s \neq 0, 1, 2$$
$$\{r \sin[p\ln(r)], r \cos[p\ln(r)]\} \times$$
$$\{\cos(\theta)\cosh(p\theta), \sin(\theta)\sinh(p\theta), \cos(\theta)\sinh(p\theta), \sin(\theta)\cosh(p\theta)\}, \quad p \neq 0, 1, 2$$

$$\{1, r^2\}\{1, \theta, \cos(2\theta), \sin(2\theta)\},$$
$$\{r, r\ln(r)\}\{\cos(\theta), \sin(\theta), \theta\cos(\theta), \theta\sin(\theta)\},$$
$$\{\cos(\mu\theta), \sin(\mu\theta)\}\{r^\mu, r^{-\mu}, r^{2-\mu}, r^{2+\mu}\},$$
$$\{\cosh(\nu\theta), \sinh(\nu\theta)\}\{\cos[\nu\ln(r)], \sin[\nu\ln(r)], r^2\cos[\nu\ln(r)], r^2\sin[\nu\ln(r)]\},$$
$$\{\cos(\theta), \sin(\theta)\}\{r, r^{-1}, r\ln(r), r^3\},$$
$$\{1, \theta\}\{1, \ln(r), r^2, r^2\ln(r)\} \tag{6.3}$$

In axisymmetric cylindrical coordinates, the Stokes equation is

$$E^4\psi = \left(\frac{\partial^2}{\partial r^2} - \frac{1}{r}\frac{\partial}{\partial r} + \frac{\partial^2}{\partial z^2}\right)^2 \psi = 0 \tag{6.4}$$

The general solution is

$$\psi = \{\cos(\alpha z), \sin(\alpha z)\}\{rI_1(\alpha r), rK_1(\alpha r), r^2I_2(\alpha r), r^2K_2(\alpha r)\}, \quad \alpha \neq 0$$
$$\{rJ_1(\beta r), rY_1(\beta r)\}\{\cosh(\beta z), \sinh(\beta z), z\cosh(\beta z), z\sinh(\beta z)\}, \quad \beta \neq 0$$
$$\{1, z\}\{1, r^2, r^2\ln(r), r^4\},$$
$$\{1, r^2\}\{1, z, z^2, z^3\} \tag{6.5}$$

Here J and Y are Bessel functions and I and K are modified Bessel functions.

In axisymmetric spherical coordinates (ϱ, θ), where ϱ is the distance from the origin and θ is the angle to the symmetry axis, let $\xi = \cos(\theta)$. The solution to

$$E^4\psi = \left(\frac{\partial^2}{\partial\varrho^2} + \frac{1-\xi^2}{\varrho^2}\frac{\partial^2}{\partial\xi^2}\right)^2 \psi = 0 \tag{6.6}$$

is

$$\psi = \{\varrho^n, \varrho^{-n+1}, \varrho^{n+2}, \varrho^{-n+3}\}\{I_n(\xi), H_n(\xi)\} \tag{6.7}$$

Here I_n, H_n are Gegenbauer functions related to the Legendre polynomials if n is integer, but in general, n may be fractional or complex.

$$I_0 = 1, \quad I_1 = -\xi, \quad I_2 = \frac{1}{2}(1-\xi^2), \quad I_3 = -\frac{1}{2}\xi(1-\xi^2)$$

$$I_n = \frac{-1}{(n-1)!} \left(\frac{d}{d\xi} \right)^{n-2} \left(\frac{\xi^2 - 1}{2} \right)^{n-1}, \quad n \geq 2 \tag{6.8}$$

$$H_0 = -\xi, \; H_1 = -1,$$

$$H_n = \frac{1}{2} I_n \ln \left(\frac{1+\xi}{1-\xi} \right) + \text{polynomials in } \xi, \quad n \geq 2 \tag{6.9}$$

Notice H_0, H_1 are not independent of I_0, I_1, and H_n is unbounded on the axis at $\xi = \pm 1$.

6.2   Eddies and Eigenvalues

Recirculating eddies may exist even for Stokes flow. Consider the flow in a channel influenced by some disturbance at one end. Let the walls be described by the normalized coordinate $y = \pm 1$. We use the middle form of Eq. (6.2)

$$\psi = \left\{ e^{ax}, e^{-ax} \right\} \{ \sin(ay), \cos(ay), y \sin(ay), y \cos(ay) \} \tag{6.10}$$

Since the stream function is symmetric in y, let

$$\psi = e^{ax} [c_1 \cos(ay) + c_2 y \sin(ay)] \tag{6.11}$$

The boundary conditions are that the velocities are zero on the walls

$$\psi(x, \pm 1) = 0, \quad \psi_y(x, \pm 1) = 0 \tag{6.12}$$

For non-trivial solutions (c_1, c_2 not both zero), Eq. (6.11) yields the characteristic equation

$$\cos(a) \sin(a) + a = 0 \tag{6.13}$$

Equation (6.13) can only be satisfied if the eigenvalue a is complex. Let

$$a = \alpha + \beta i \tag{6.14}$$

where α, β are real. Substituting into Eq. (6.13) and equating real and imaginary parts give

$$\sin(2\alpha) \cosh(2\beta) + 2\alpha = 0, \quad \cos(2\alpha) \sinh(2\beta) + 2\beta = 0 \tag{6.15}$$

Using a numerical root finder yields the first five eigenvalues listed in Table 6.1.

Let the disturbance be some prescribed velocity at $x = 0$ and decaying to zero as x increases. Thus α must be negative. Using the sign change in β, one can limit to the real

Table 6.1 Real and imaginary parts of the first five eigenvalues

$\pm\alpha$	2.106	5.356	8.537	11.699	14.854
$\pm\beta$	1.125	1.552	1.776	1.929	2.047

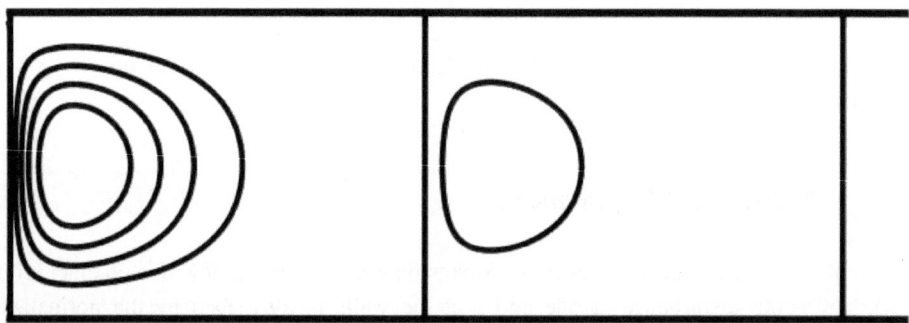

Fig. 6.1   Streamlines of Eq. (6.16), $c = 1$. Each cell has a width of π/β. Magnitude of the first cell is 0.132, while that of the second cell is 0.00037

part of the eigenfunction in Eq. (6.11).

$$\psi = ce^{\alpha x} \sin(\beta x)\{ \cosh(\beta) \sinh(\beta y)[\sin(\alpha) \sin(\alpha y) + \cos(\alpha)y\cos(\alpha y)]$$
$$- \sinh(\beta)\cosh(\beta y)[\cos(\alpha) \cos(\alpha y) + \sin(\alpha)y \sin(\alpha y)]\} \tag{6.16}$$

Figure 6.1 shows the stream function for the first eigenvalue ($\alpha = -2.106$, $\beta = 1.125$). Notice there is a string of recirculating eddies (Moffatt eddies). However, the strength of the eddies decreases exponentially, such that any physical significance (e.g. transport properties) of the weaker eddies are almost nil.

6.3   The Scraping Problem

In the scraping problem, a bottom plate moves with velocity U and a fixed top plate is slanted at an angle β. The high Reynolds numbers case was considered in Sect. 4.4.1 (Fig. 4.8). In this section, the low Reynolds number Stokes scraping problem is studied. Since the velocity is constant, the stream function is proportional to r. Using the fourth form of Eq. (6.3), the stream function, which is zero on $\theta = 0$, is

$$\psi = Ur[c_1 \sin(\theta) + c_2\theta \cos(\theta) + c_3\theta \sin(\theta)] \tag{6.17}$$

The other boundary conditions

Fig. 6.2 Streamlines for
scraping at an angle

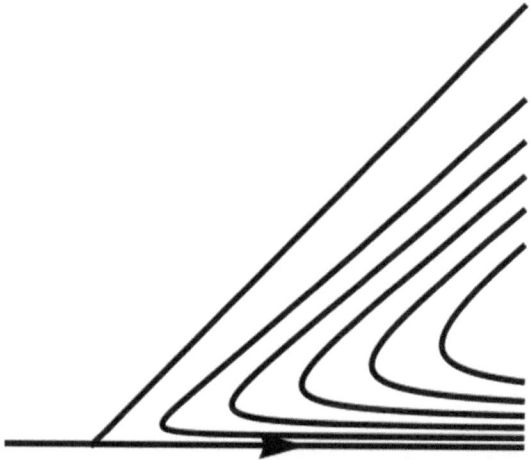

$$\psi(r,\beta) = 0, \quad \psi_\theta(r,\beta) = 0, \quad \frac{\psi_\theta(r,0)}{r} = U \tag{6.18}$$

yield

$$c_1 = \frac{2\beta^2}{2\beta^2 + \cos(2\beta) - 1}, \quad c_2 = \frac{-2\sin^2(\beta)}{2\beta^2 + \cos(2\beta) - 1}, \quad c_3 = \frac{-2\beta + \sin(2\beta)}{2\beta^2 + \cos(2\beta) - 1}$$
$$\tag{6.19}$$

Figure 6.2 shows the streamlines. Notice the velocity is slower (streamlines more separated) near the fixed plate. Since the Stokes equation is linear, the direction of the velocity
can be reversed.

6.4   Flow Through Tapered Channels and Cones

Figure 6.3a shows a divergent channel of opening angle 2β and fluid is supplied at the
vertex with rate Q. The exact N-S solution is the Jeffery-Hamel flow (Sect. 2.2.4), using
complicated elliptic integrals. The Stokes flow solution is simpler. Normalize the stream
function by $Q/2$ and assume radial flow. From Eq. (6.3), the stream function is a linear
combination of $\{1, \theta, \theta^2, \theta^3\}$. The boundary conditions are that on $\theta = \pm\beta$, $\psi = \pm 1$
and $\psi_\theta = 0$. The solution is

$$\psi = \frac{3\theta}{2\beta} - \frac{\theta^3}{2\beta^3} \tag{6.20}$$

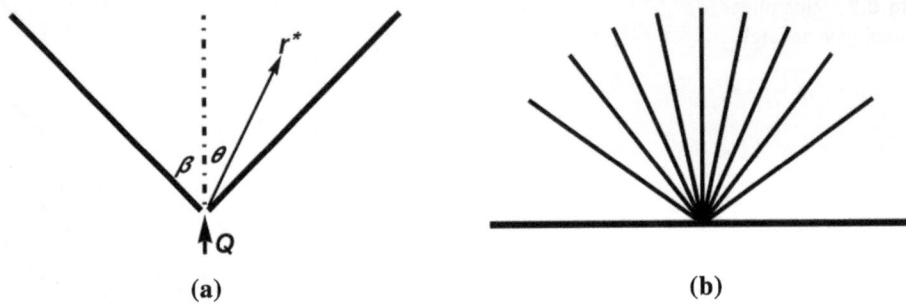

Fig. 6.3 **a** Fluid is supplied at the vertex of a tapered channel **b** The streamlines emanating from a line source on a plane. $\beta = \frac{\pi}{2}$, $\Delta\psi = 0.2$

For the flow through a slit on a plane, set $\beta = \pi/2$. This solution is plotted in Fig. 6.3b. Notice the spacing of the streamlines is uneven, which differs from a two-dimensional potential source. Due to the linearity of the Stokes equation, the direction of the flow is reversible. Also two different sources on the same plane can be superposed.

Next, consider the flow through a cone (rotate Fig. 6.3a about the symmetry axis). In this case, there is no exact N-S solution. The general solution of the stream function in spherical coordinates, normalized by $Q/(2\pi)$, is given by Eq. (6.7). We discard $H_n(\xi)$ which is unbounded on the axis. Thus, the solution independent of the radial coordinate is

$$\psi = c_1 I_0 + c_2 I_1 + c_3 I_3 = c_1 - c_2 \xi - \frac{1}{2} c_3 \xi \left(1 - \xi^2\right) \tag{6.21}$$

where $\xi = \cos(\theta)$. The boundary conditions are that on $\theta = \beta$, $\psi = 1$ and $\psi_\xi = 0$, and that on $\theta = 0$, $\psi = 0$. The solution is

$$\psi = \frac{(1 - \xi)}{(1 + 2\xi_0)(1 - \xi_0)^2} \left(1 - 3\xi_0^2 + \xi + \xi^2\right) \tag{6.22}$$

Here $\xi_0 = \cos(\beta)$. In the case of fluid emerging through a hole on a plate, take $\beta = \pi/2$ and Eq. (6.22) becomes

$$\psi = 1 - \xi^3 = 1 - \cos^3(\theta) \tag{6.23}$$

The streamlines are similar to Fig. 6.3b.

6.5 Flow Due to Rotating Disks

Figure 6.4 shows a fixed disk over a rotating disk with rotation speed Ω. Assume the disk radii are much larger the gap width a. Use axisymmetric cylindrical coordinates and normalize with a and Ω. For Stokes flow we expect only the azimuthal velocity v in Eq. (1.40) remains. The Stokes equation for rotation is

$$\frac{\partial^2 v}{\partial r^2} + \frac{1}{r}\frac{\partial v}{\partial r} + \frac{\partial^2 v}{\partial z^2} - \frac{v}{r^2} = 0 \tag{6.24}$$

with the boundary conditions $v(r, 0) = r$, $v(r, 1) = 0$. The solution using separation of variables is

$$v = r(1 - z) \tag{6.25}$$

and is shown in Fig. 6.4. These are constant velocity lines, not streamlines. If the top disk is rotating with a different speed $\alpha\Omega$, then Eq. (6.25) can be superposed with $v = \alpha r z$.

6.6 The Translating Sphere

The Stokes drag of a sphere is important for the motion or settling of small particles. Consider a sphere with diameter a placed in a uniform stream of velocity U. Normalize the stream function by Ua^2. The Stokes equation and its general solution in spherical coordinates are given by Eqs. (6.6, 6.7). Equations (1.52) show the stream function for uniform flow at infinity is $\varrho^2 \sin^2\theta/2$. Since $\sin^2\theta = 1 - \xi^2$, the solution is

$$\psi = \{\varrho^2, \varrho^{-1}, \varrho^4, \varrho\}I_2 = \frac{1}{2}(1 - \xi^2)\left(c_1\varrho^2 + \frac{c_2}{\varrho} + c_3\varrho\right) \tag{6.26}$$

where we have discarded the unbounded H_2 and ϱ^4. The boundary condition, that ψ tend to uniform flow at infinity, gives $c_1 = 1$, while the no-slip conditions on $\varrho = 1$ yield

$$\psi = \frac{1}{2}\sin^2\theta\left(\varrho^2 + \frac{1}{2\varrho} - \frac{3}{2}\varrho\right) \tag{6.27}$$

Fig. 6.4 Azimuthal constant velocity lines for a fixed disc over a rotating disk

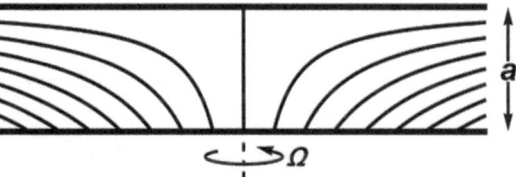

Fig. 6.5 Stokes flow over a
sphere. Streamlines from the
sphere: $\psi = 0, 0.05, 0.2, 0.4,$
$0.6, 0.8, 1$

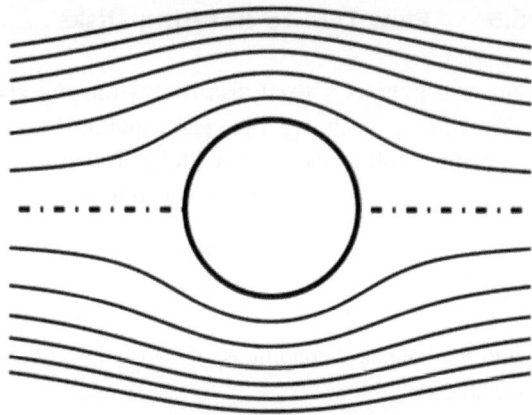

The streamlines are shown in Fig. 6.5.

The drag on the sphere is the sum of pressure drag and shear drag. Using Eq. (1.52) and Eqs. (1.47, 1.48), the Stokes pressure can be shown to be

$$\frac{\partial p}{\partial \varrho} = \frac{1}{\varrho^2 \sin(\theta)} \frac{\partial}{\partial \theta} E^2 \psi, \quad \frac{1}{\varrho} \frac{\partial p}{\partial \theta} = \frac{-1}{\varrho \sin(\theta)} \frac{\partial}{\partial \varrho} E^2 \psi \tag{6.28}$$

where p is the pressure normalized by $\frac{\mu U}{a}$, and $\mu = \rho \nu$ is the viscosity. Upon substitution of Eq. (6.27) yields

$$p = p_\infty - \frac{3 \cos(\theta)}{2\varrho} \tag{6.29}$$

Here p_∞ is the pressure at infinity. The drag due to pressure is integrated over the sphere surface

$$D_p = -\int\limits_0^\pi p 2\pi \sin(\theta) \cos(\theta) d\theta = 2\pi \tag{6.30}$$

Using Eq. (1.51), the normalized shear stress on the surface is

$$\tau = 2d_{\varrho\theta}\big|_{r=1} = \frac{\partial \upsilon}{\partial \varrho}\bigg|_{r=1} = \frac{-1}{\sin(\theta)} \left(\frac{1}{\varrho}\psi_\varrho\right)_\varrho = -\frac{3}{2}\sin(\theta) \tag{6.31}$$

The drag due to shear is

$$D_s = -\int\limits_0^\pi \tau 2\pi \sin^2(\theta) d\theta = 4\pi \tag{6.32}$$

The total drag, normalized by $a\mu U$, is

$$D = 2\pi + 4\pi = 6\pi \tag{6.33}$$

Notice both pressure and shear contribute significantly to the total drag. By using appropriate coordinate systems, the drag of other three-dimensional bodies can also be derived.

6.7   The Translating Cylinder and the Stokes Paradox

Consider two-dimensional uniform Stokes flow over a cylinder. Normalize by the free stream velocity U and the radius of the cylinder a. In cylindrical coordinates (r, θ), the stream function for uniform flow at infinity is $\psi = r\sin(\theta)$. Equation (6.3) shows the solution is a linear combination of

$$\psi = \sin(\theta)\{r, r^{-1}, r\ln(r), r^3\} \tag{6.34}$$

The boundary conditions are that ψ tends to $r\sin(\theta)$ at infinity, and that there is no-slip on the cylinder at $r = 1$, i.e. $\psi = 0$, $\psi_r = 0$. But Eq. (6.34) could not satisfy all these conditions and we have Stokes paradox.

The non-existence of (steady) Stokes flow over two-dimensional bodies is usually attributed to the neglected convection terms, but a better explanation is through dimensional analysis (e.g. Happel and Brenner 1973). For the flow over a cylinder, the parameters are drag per length D [M/T²], viscosity μ [M/LT], velocity U [L/T] and radius a [L]. Section 3.1 shows there is only one non-dimensional group left

$$\frac{D}{\mu U} = \text{constant} \tag{6.35}$$

Thus, the drag is independent of radius a, which could not be true. For the sphere, drag D' is force with dimension [ML/T²] and, from the previous section,

$$\frac{D'}{a\mu U} = 6\pi \tag{6.36}$$

which is consistent.

Now could we add another length scale to the two-dimensional problem so that the radius enters the Pi theorem with a nondimensional ratio of two lengths? This depends on what length scale is introduced. For example, an elliptic cylinder has two length scales, but the paradox persists. Adding a streamwise parallel infinite plate also does not help.

The fact is that the Stokes paradox can be removed if the domain containing the cylinder is bounded by a finite transverse distance to the oncoming flow. For example, the solution exists when Stokes flow passes through a perpendicular screen composed of an infinite row of cylinders (domain of each cylinder is transversely bounded). However, there is no solution if the plane of the screen is parallel to the flow.

An example for avoiding the Stokes paradox is the Happel cell. Consider a concentric larger cylinder of radius $b > 1$ limiting the domain. On the cell, the velocity is prescribed to be uniform parallel flow, i.e. $\psi = r\sin(\theta)$. Now Eq. (6.34) is

$$\psi = \sin(\theta) f(r), \quad f = c_1 r + \frac{c_2}{r} + c_3 r \ln(r) + c_4 r^3 \tag{6.37}$$

The boundary conditions are that on $r = b$

$$u(b, \theta) = \left(\frac{1}{r}\frac{\partial \psi}{\partial \theta}\right)(b, \theta) = \cos(\theta) \tag{6.38}$$

$$v(b, \theta) = \left(-\frac{\partial \psi}{\partial r}\right)(b, \theta) = -\sin(\theta) \tag{6.39}$$

and on $r = 1$

$$u(1, \theta) = 0, \quad v(1, \theta) = 0 \tag{6.40}$$

The four conditions determine the four constants

$$c_1 = \frac{1 - b^2}{A}, \quad c_2 = \frac{b^2}{A}, \quad c_3 = \frac{2(1 + b^2)}{A}, \quad c_4 = \frac{-1}{A}$$
$$A = 2\left[1 - b^2 + \left(1 + b^2\right)\ln(b)\right] \tag{6.41}$$

Fig. 6.6 Streamlines in a Happel cell with $b = 2$

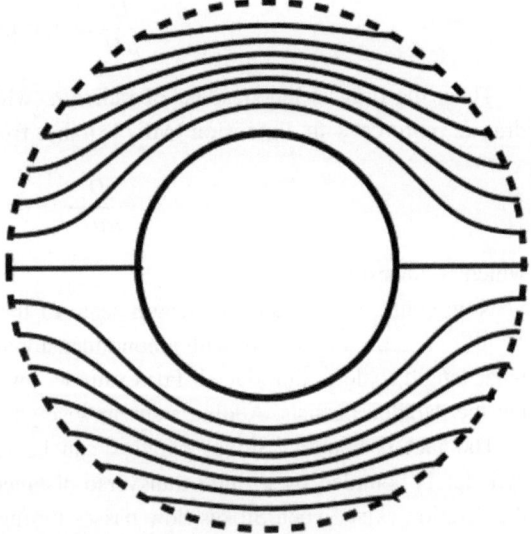

Figure 6.6 shows the streamlines. Since the domain is bounded (especially vertically), Stokes paradox is averted. The cell model is sometimes used for the flow through a two-dimensional array of parallel cylinders.

6.8   Unsteady Stokes Flow

As in Chap. 5, we shall concentrate on Stokes flows where unsteadiness is important and exclude parallel or concentric unsteady flows which have exact solutions.

For axial rotation of an axisymmetric body, the streamlines are circles about the axis. The unsteady Stokes equation is a diffusion equation in the azimuthal velocity. For other fluid motions for which a stream function exists, the unsteady Stokes equation is of the form

$$\frac{\partial}{\partial t} L^2 \psi = L^4 \psi \tag{6.42}$$

where L^2 is some second order differential linear space operator, which could be ∇^2 or E^2. Since L^2 and the time derivative commute, the solution is a linear combination of the solutions of

$$L^2 \psi = 0 \tag{6.43}$$

and

$$\frac{\partial}{\partial t} \psi = L^2 \psi \tag{6.44}$$

provided all four solutions are independent. Equation (6.43) yields quasi-steady solutions, i.e., they can be multiplied by any function of time. These solutions are represented by two of the independent solutions of steady Stokes flow given in Sect. 6.1. Equation (6.44) is a diffusion equation. It can be solved by separation of variables, stretching transform (Appendix A), or by Laplace transform. The stretching transform gives similarity solutions which are related to squeeze flows. The Laplace transform method eliminates the time derivative easily, but the inverse transform is complicated for any unsteady motion other than constant velocity impulsive start.

The following are some representative examples which leads to closed-form solutions by using separation of variables. Since Stokes flow is linear, it is possible to superpose unsteady motions. Thus, a Green's function solution in integral form can be constructed for general unsteady motion.

6.8.1 Unsteady Rotation of an Infinite Disk

For rotating Stokes flow, as in Sect. 6.5, only the azimuthal velocity component exists. Consider a single disk rotating in its own plane with angular velocity $\Omega_0 h(t)$ where the time has been normalized by $1/\Omega_0$ and $h(0) = 1$. Equation (1.45) gives the normalized unsteady Stokes equation in cylindrical coordinates

$$\frac{\partial}{\partial t}(rv) = E^2(rv), \quad E^2 = \frac{\partial^2}{\partial r^2} - \frac{1}{r}\frac{\partial}{\partial r} + \frac{\partial^2}{\partial z^2} \tag{6.45}$$

where v is the azimuthal velocity. Since on the disk $v = r$, separation of variables yields

$$v = re^{\lambda t} f(z) \tag{6.46}$$

Equation (6.45) gives

$$f'' - \lambda f = 0 \tag{6.47}$$

The boundary conditions are

$$f(0) = 1$$

and that f decays to zero at infinity. This is possible only if $\lambda > 0$. The solution is

$$f = e^{-\sqrt{\lambda}z} \tag{6.48}$$

Equation (6.48) is plotted in Fig. 6.7. Since λ is positive, the rotation is accelerated and does not include the steady rotation case. Torsional oscillation about the axis is possible (see Exercise 6.5).

6.8.2 Accelerated Rotating Sphere

Let the radius be a and the angular velocity about a diameter be $\Omega_0 h(t)$ as in the previous section. The normalized governing equation from Eq. (1.54), in spherical coordinates, is

Fig. 6.7 Constant velocity lines of an accelerated rotation of an infinite disk at a given time

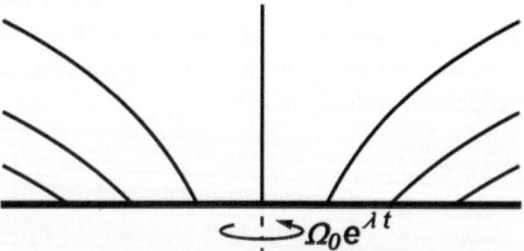

$$\frac{\partial}{\partial t}\chi = E^2\chi, \quad E^2 = \frac{\partial^2}{\partial\varrho^2} + \frac{1}{\varrho^2}\frac{\partial^2}{\partial\theta^2} - \frac{\cot(\theta)}{\varrho^2}\frac{\partial}{\partial\theta} \tag{6.49}$$

and χ is related to the azimuthal velocity w by

$$\chi = \varrho \sin(\theta)w \tag{6.50}$$

On the surface of the sphere $\varrho = 1$

$$w = \sin(\theta)h(t) \tag{6.51}$$

Thus, set

$$\chi = e^{\lambda t}\sin^2(\theta)f(\varrho) \tag{6.52}$$

where we let $h(t) = e^{\lambda t}$ in order to separate the variables. Equation (6.49) gives

$$\lambda f = f'' - \frac{2}{\varrho^2}f \tag{6.53}$$

with the boundary conditions

$$f(1) = 1, \quad f(\infty) = 0 \tag{6.54}$$

The only solution is when $\lambda = b^2 \geq 0$, or the sphere is steady or accelerating,

$$f = \frac{\left(b + \frac{1}{\varrho}\right)}{(b+1)}e^{-b(\varrho-1)} \tag{6.55}$$

Thus, the velocity is

$$w = e^{b^2 t}\sin(\theta)\frac{\left(b + \frac{1}{\varrho}\right)}{\varrho(b+1)}e^{-b(\varrho-1)} \tag{6.56}$$

For steady rotation, set $b = 0$

$$w = \frac{\sin(\theta)}{\varrho^2} \tag{6.57}$$

Typical constant velocity lines are shown in Fig. 6.8. We find Stokes solutions for decelerated rotation do not exist.

However, torsional oscillation about the axis is possible. Let $b^2 = i\omega$, where ω is the frequency. Then replace

$$b = \frac{(1+i)}{\sqrt{2}}\sqrt{\omega} \tag{6.58}$$

Fig. 6.8 Constant velocity
lines for the steady rotation of
a sphere

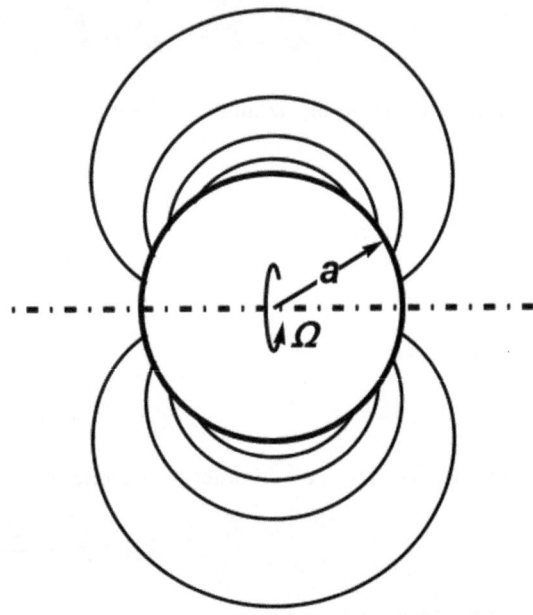

in Eqs. (6.56), keeping in mind that only the real part of the product has physical
significance. Spatial oscillations in the radial direction occur.

6.8.3   Unsteady Translation of a Sphere

Normalize with the sphere radius a and the infinity velocity at $t = 0$. The governing
equation is Eq. (6.42) with

$$L^2 = E^2 = \frac{\partial^2}{\partial \varrho^2} + \frac{1}{\varrho^2}\frac{\partial^2}{\partial \theta^2} - \frac{\cot(\theta)}{\varrho^2}\frac{\partial}{\partial \theta} \tag{6.59}$$

The unsteady uniform stream function at infinity is $e^{\lambda t}\varrho^2 \sin^2\theta/2$ and we assumed an
exponential time dependence. Let

$$\psi = e^{\lambda t}\frac{\sin^2\theta}{2}f(\varrho) \tag{6.60}$$

Equation (6.43) gives

$$f = c_1\varrho^2 + \frac{c_2}{\varrho} \tag{6.61}$$

Equation (6.44) gives the bounded solution

$$f = c_3\left(b + \frac{1}{\varrho}\right)e^{-b\varrho}, \quad b = \sqrt{\lambda} > 0 \tag{6.62}$$

(The other solution, Eq. (6.62) with b replaced by $-b$, is unbounded). The complete solution is the sum of Eqs. (6.61, 6.62). The boundary conditions are

$$f(1) = 0, \quad f'(1) = 0, \quad f(\infty) \sim \varrho^2 \tag{6.63}$$

Thus

$$c_1 = 1, \quad c_2 = \frac{-(b^2 + 3b + 3)}{b^2}, \quad c_3 = \frac{3e^b}{b^2} \tag{6.64}$$

and

$$\psi = e^{b^2 t}\frac{\sin^2\theta}{2}\left[\varrho^2 + \frac{c_2}{\varrho} + c_3 e^{-b\varrho}\left(b + \frac{1}{\varrho}\right)\right] \tag{6.65}$$

For steady flow, $b = 0$ and the last two solutions in Eq. (6.65) are not independent. The stream function is then replaced by Eq. (6.27).

If the sphere is oscillating along a diameter, substitute Eq. (6.58) into Eq. (6.64, 6.65).

6.8.4   Unsteady Translation of a Cylinder

Consider Stokes flow due to a circular cylinder moving normal to its axis. Section 6.7 shows steady flow does not exist (Stokes paradox). Let us investigate whether unsteady flow exists.

Use cylindrical coordinates fixed on the cylinder and normalize with the radius a and infinity velocity at time zero. The governing equation is Eq. (6.42) with

$$L^2 = \nabla^2 = \frac{\partial^2}{\partial r^2} + \frac{1}{r}\frac{\partial}{\partial r} + \frac{1}{r^2}\frac{\partial^2}{\partial \theta^2} \tag{6.66}$$

The stream function should approach unsteady uniform flow at infinity

$$\psi \sim e^{\lambda t}\sin(\theta)r \tag{6.67}$$

Let

$$\psi = e^{\lambda t}\sin(\theta)f(r) \tag{6.68}$$

Equation (6.43) gives the solution

$$f = c_1 r + \frac{c_2}{r} \tag{6.69}$$

Equation (6.44) yields

$$f = c_3 K_1\left(\sqrt{\lambda} r\right) + c_4 I_1\left(\sqrt{\lambda} r\right), \quad \lambda > 0 \tag{6.70}$$

$$f = c_5 J_1\left(\sqrt{|\lambda|} r\right) + c_6 Y_1\left(\sqrt{|\lambda|} r\right), \quad \lambda < 0 \tag{6.71}$$

Here K_1, I_1 are modified Bessel functions and J_1, Y_1 are Bessel functions. If $\lambda = 0$ or steady flow, the solutions to Eq. (6.44) are not independent of those of Eq. (6.43), and we have the solution Eq. (6.37), leading to Stokes paradox. For unsteady flow, the complete solution is a combination of Eq. (6.69) and Eq. (6.70) or Eq. (6.71). The boundary conditions are

$$f(1) = 0, \quad f'(1) = 0, \quad f(\infty) \sim r \tag{6.72}$$

Now if $\lambda > 0$ or accelerating flow, c_4 is set to zero since the velocity is unbounded, and the solution is

$$\psi = e^{\lambda t} \sin(\theta)\left[r + \frac{c_2}{r} + c_3 K_1\left(\sqrt{\lambda} r\right)\right], \quad c_2 = -\frac{K_2\left(\sqrt{\lambda}\right)}{K_0\left(\sqrt{\lambda}\right)}, \quad c_3 = \frac{2}{\sqrt{\lambda} K_0\left(\sqrt{\lambda}\right)} \tag{6.73}$$

This flow is shown in Fig. 6.9. The corresponding pressure drag and shear drag can also be found. However, the solution ceased to exist when we approach the steady-state limit $\lambda \to 0$.

Fig. 6.9 Streamlines of an accelerating circular cylinder

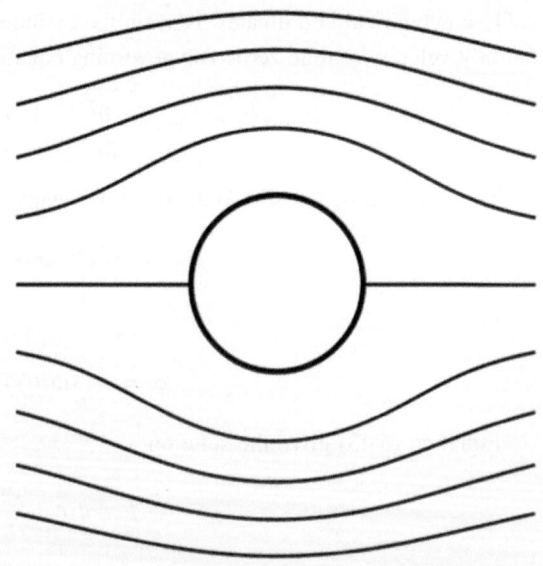

For decelerating flow, Eq. (6.71) shows both J_1, Y_1 are oscillatory, but both decay to zero far from the cylinder. Since the coefficients cannot be uniquely determined, the conclusion is that the decelerating Stokes flow solution does not exist. Physically, for decelerating flow, the solution is so sensitive to the past history such that there is no unique solution.

For a cylinder oscillating along a diameter the solution is

$$\psi = e^{i\omega t} \sin(\theta)\left[r + \frac{c_2}{r} + c_3 g(r)\right] \tag{6.74}$$

where

$$g = \ker_1\left(\sqrt{\omega}r\right) + i\kei_1\left(\sqrt{\omega}r\right), \quad c_2 = -1 - g(1), \quad c_3 = \frac{-[2 + g(1)]}{g'(1)} \tag{6.75}$$

and ker, kei are Kelvin functions.

6.9 Narrow Gap Approximation

A further approximation for Stokes flow can be made when the fluid domain is narrow, such as that in a crevice.

6.9.1 Steady Flow

Figure 6.10 shows a fluid is contained between a two-dimensional gap of length L and a maximum height of εL where $\varepsilon \ll 1$. The bottom plate is flat and moving with constant velocity U and the top boundary is fixed at $y^* = \varepsilon L h(x^*)$. Normalize all lengths by L, the velocities by U, the pressure by $\mu U/\varepsilon^2 L$ and the stream function by $\varepsilon U L$. Let

$$y = \varepsilon \eta \tag{6.76}$$

The Laplacian operator becomes

$$\nabla^2 = \frac{\partial^2}{\partial x^2} + \frac{\partial^2}{\varepsilon^2 \partial \eta^2} \tag{6.77}$$

From Eq. (1.16), the normalized Stokes equation gives the approximation

Fig. 6.10 Fluid is in the gap between the fixed top and the moving bottom plate

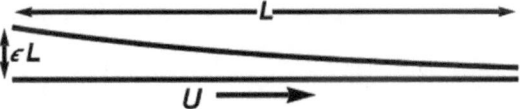

$$p_x = u_{\eta\eta} = \psi_{\eta\eta\eta} \tag{6.78}$$

Equation (1.17) yields

$$p_\eta = 0 \tag{6.79}$$

As in a boundary layer, the pressure is constant across the gap. Equations (6.78, 6.79) lead to the Stokes equation

$$\psi_{\eta\eta\eta\eta} = 0 \tag{6.80}$$

The boundary conditions are

$$\psi|_{\eta=0} = 0, \quad \psi_\eta|_{\eta=0} = 1, \quad \psi|_{\eta=h} = c, \quad \psi_\eta|_{\eta=h} = 0 \tag{6.81}$$

Here $h(x)$ is the given shape of the top boundary, c is a constant representing normalized mass flux. The solution is

$$\psi = \eta + c_2(x)\eta^2 + c_3(x)\eta^3, \quad c_2 = \frac{1}{h^2}(3c - 2h), \quad c_3 = \frac{-1}{h^3}(2c - h) \tag{6.82}$$

Equation (6.78) gives the pressure gradient

$$p_x = \frac{-12c}{h^3} + \frac{6}{h^2} \tag{6.83}$$

This leads to the Reynolds' equation (see e.g. Sherman 1990) used extensively in lubrication theory

$$\frac{d}{dx}(h^3 p_x) = 6\frac{dh}{dx} \tag{6.84}$$

Equation (6.83) can be integrated

$$p(x) = p_0 - 12c \int_0^x \frac{1}{h^3}dx + 6\int_0^x \frac{1}{h^2}dx \tag{6.85}$$

If at the exit the pressure is also p_0, the constant is

$$c = \frac{\int_0^1 \frac{1}{h^2}dx}{2\int_0^1 \frac{1}{h^3}dx} \tag{6.86}$$

The pressure distribution can be integrated exactly for a variety of the crevice shapes such as $h = \cos(\beta x)$, $e^{-\beta x}$, $(1 - \beta x)^\gamma$. The lifting force on the top surface is

$$F = \int_0^1 [p(x) - p_0]dx \tag{6.87}$$

6.9.2   Unsteady Flow

In Sect. 5.3 we studied the squeezing between two plates with small squeeze number (unsteadiness not quite important). In this section the unsteadiness is as important as the viscous terms, but with a narrow gap approximation. Although the unsteady Stokes equation, Eq. (6.42) is linear, in order to obtain an analytic solution, the rate of squeezing is prescribed.

Let the top plate be squeezing down with a gap height $ah(t)$, where a is the initial distance between the plates. Normalize with length scale a and time scale T. Using the narrow gap assumption, the unsteady Stokes equation, Eq. (6.42), becomes

$$\psi_{\eta\eta t} = \psi_{\eta\eta\eta\eta} \tag{6.88}$$

The solution is a combination of the solutions of

$$\psi_{\eta\eta} = 0 \tag{6.89}$$

$$\psi_t = \psi_{\eta\eta} \tag{6.90}$$

Let $h(t) = \sqrt{1 - \alpha t}$, where $\alpha > 0$. A transform similar to Eq. (2.27) is

$$\psi = \frac{x}{\sqrt{1 - \alpha t}} f(\zeta), \quad \zeta = \frac{\eta}{\sqrt{1 - \alpha t}} \tag{6.91}$$

The independent solutions to Eq. (6.89) are 1, ζ, while Eq. (6.90) gives

$$\frac{\alpha}{2}(f + \zeta f') = f'' \tag{6.92}$$

If the bottom plate is still, the boundary conditions are

$$f(0) = 0, \quad f'(0) = 0 \tag{6.93}$$

On the top plate at $\zeta = 1$, the horizontal velocity is zero, while the vertical velocity is

$$v = -\psi_x = \frac{dh}{dt} \tag{6.94}$$

or

$$f'(1) = 0, \quad f(1) = \frac{\alpha}{2} \tag{6.95}$$

The general solution is

$$f = c_0 + c_1\zeta + c_2 e^{\alpha\zeta^2/4} + c_3 e^{\alpha\zeta^2/4}\mathrm{erf}\left[\sqrt{\alpha}\frac{\zeta}{2}\right] \tag{6.96}$$

Fig. 6.11 Instantaneous streamlines for squeezing from the top plate

where the last two independent solutions are from Eq. (6.92), and erf is the error function. The boundary conditions give

$$c_0 = \frac{\alpha\sqrt{\pi}\,\mathrm{erf}\left(\frac{\sqrt{\alpha}}{2}\right)}{\Delta}, \quad c_1 = \frac{-\alpha^{\frac{3}{2}}}{\Delta}, \quad c_2 = \frac{-\alpha\sqrt{\pi}\,\mathrm{erf}\left(\frac{\sqrt{\alpha}}{2}\right)}{\Delta}, \quad c_4 = \frac{\alpha\sqrt{\pi}}{\Delta},$$

$$\Delta = 2\left[\sqrt{\pi}\,\mathrm{erf}\left(\frac{\sqrt{\alpha}}{2}\right) - \sqrt{\alpha}\right] \tag{6.97}$$

Typical streamlines are shown in Fig. 6.11.

6.10   Boundary Perturbation

The boundary conditions are difficult to apply if a surface is striated or wavy, such as those due to milling or roughness. An analytic solution may be possible by perturbation using the small amplitude of the striations.

6.10.1 Couette Flow over a Wavy Fixed Plate

Figure 6.12 shows shearing (Couette) flow. The top plate is moving in its own plane, with velocity U at a distance a from the wavy bottom plate described by

$$y^* = b\sin\left(\frac{2\pi x^*}{l}\right) \tag{6.98}$$

Normalize velocities by U and lengths by a. Equation (6.98) becomes

$$y = \varepsilon\sin(\lambda x) \tag{6.99}$$

Fig. 6.12 Couette flow over a wavy bottom

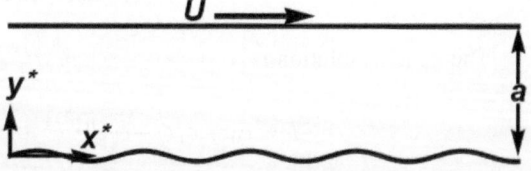

where $\varepsilon = \frac{b}{a} \ll 1$ and $\lambda = \frac{2\pi a}{l} = O(1)$. The Stokes equation is the biharmonic

$$\nabla^4 \psi = \left(\frac{\partial^2}{\partial x^2} + \frac{\partial^2}{\partial y^2}\right)^2 \psi = 0 \tag{6.100}$$

The boundary conditions are

$$\psi_x|_{y=1} = 0, \quad \psi_y|_{y=1} = 1 \tag{6.101}$$

$$\psi_x|_{y=\varepsilon \sin(\lambda x)} = 0, \quad \psi_y|_{y=\varepsilon \sin(\lambda x)} = 0 \tag{6.102}$$

Equations (6.102) is then expanded using Taylor series, e.g.,

$$\psi_y|_{y=\varepsilon \sin(\lambda x)} = \psi_y|_{y=0} + \varepsilon \sin(\lambda x)\,\psi_{yy}|_{y=0} + \frac{\varepsilon^2 \sin^2(\lambda x)}{2!}\,\psi_{yyy}|_{y=0} + \cdots \tag{6.103}$$

Expand the stream function in terms of the small parameter ε (see Appendix B)

$$\psi = \psi_0 + \varepsilon \psi_1 + \varepsilon^2 \psi_2 + \cdots \tag{6.104}$$

The zeroth order is

$$\nabla^4 \psi_0 = 0, \quad \psi_{0x}(x,1) = 0, \quad \psi_{0y}(x,1) = 1, \quad \psi_{0x}(x,0) = 0, \quad \psi_{0y}(x,0) = 0 \tag{6.105}$$

The solution is the Couette flow for smooth plates

$$\psi_0 = y^2/2 \tag{6.106}$$

The first order is

$$\nabla^4 \psi_1 = 0, \quad \psi_{1x}(x,1) = 0, \quad \psi_{1y}(x,1) = 0, \quad \psi_{1x}(x,0) = -\sin(\lambda x)\psi_{0xy}(x,0) = 0$$
$$\psi_{1y}(x,0) = -\sin(\lambda x)\psi_{0yy}(x,0) = -\sin(\lambda x) \tag{6.107}$$

The solution is

$$\psi_1 = \sin(\lambda x)\{c_1 \langle \sinh[\lambda(1-y)] - \lambda(1-y)\cosh[\lambda(1-y)]\rangle + c_2(1-y)\sinh[\lambda(1-y)]\}$$
$$c_1 = \frac{2\sinh(\lambda)}{1 + 2\lambda^2 - \cosh(2\lambda)}, \quad c_2 = \frac{2[\lambda \cosh(\lambda) - \sinh(\lambda)]}{1 + 2\lambda^2 - \cosh(2\lambda)} \tag{6.108}$$

We shall leave the second order solution as an exercise.

6.10.2 Flow Through a Channel with Bumpy Walls

Laminar flow between parallel flat plates leads to the well-known Poisueille parabolic velocity profile. What if the walls are not flat but have three-dimensional bumps (Fig. 6.13)?. This problem is one of the rare occasions where the primary variables (not the stream function) should be used. Let z be the transverse coordinate and the mean pressure gradient is along the x direction. Normalizing as in the previous section, the walls are described by

$$z = 1 + \varepsilon \sin(\alpha x) \sin(\beta y), \quad z = \pm \varepsilon \sin(\alpha x) \sin(\beta y) \qquad (6.109)$$

where the $\pm$ sign indicates the bumps are in-phase or out-of-phase. Let the velocities be (u, v, w) in the directions (x, y, z) respectively. The continuity and Stokes equations are

$$u_x + v_y + w_z = 0 \qquad (6.110)$$

$$p_x = u_{xx} + u_{yy} + u_{zz}, \quad p_y = v_{xx} + v_{yy} + v_{zz}, \quad p_z = w_{xx} + w_{yy} + w_{zz} \qquad (6.111)$$

Next, we perturb the primary variables in ε, keeping in mind the boundary conditions should be expanded in a Taylor series as in Eq. (6.103). The zeroth order is Poiseuille flow

$$u_0 = \frac{z(1-z)}{2}, \quad v_0 = 0, \quad w_0 = 0, \quad p_0 = -x \qquad (6.112)$$

The boundary conditions for u_1 are

$$u_1|_{z=1} = -\sin(\alpha x)\sin(\beta y)\, u_{0z}|_{z=1}, \quad u_1|_{z=0} = \mp \sin(\alpha x)\sin(\beta y)\, u_{0z}|_{z=0} \qquad (6.113)$$

Fig. 6.13 Flow between two bumpy plates

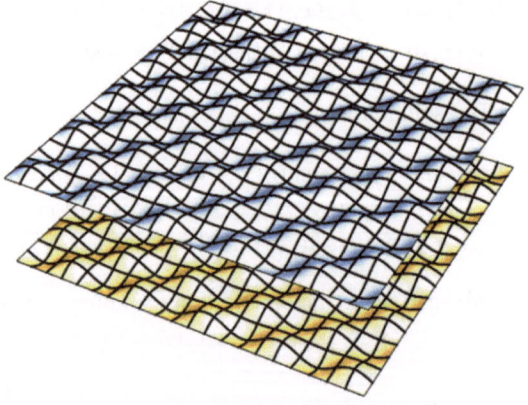

Equations (6.113, 6.110, 6.111) suggest

$$u_1 = \sin(\alpha x)\sin(\beta y)U(z), \quad v_1 = \cos(\alpha x)\cos(\beta y)V(z),$$
$$w_1 = \cos(\alpha x)\sin(\beta y)W(z), \quad p_1 = \cos(\alpha x)\sin(\beta y)P(z) \tag{6.114}$$

Thus, Eqs. (6.110, 6.111) reduce to the ordinary differential equations

$$\alpha U - \beta V + W' = 0, \quad -\alpha P = U'' - (\alpha^2 + \beta^2)U,$$
$$\beta P = V'' - (\alpha^2 + \beta^2)V, \quad P\prime = W'' - (\alpha^2 + \beta^2)W \tag{6.115}$$

Eliminate P and solve for U, V, W, using the boundary conditions

$$U(1) = \frac{1}{2}, \quad V(1) = 0, \quad w(1) = 0, \quad U(0) = \mp\frac{1}{2}, \quad V(0) = 0, \quad W(0) = 0 \tag{6.116}$$

The solution to Eqs. (6.115, 6.116), not presented here, can be obtained as in Wang (2004).

6.11 Eigenfunction Expansions

For some problems, no simplifying approximations can be made. If the boundary can be described by separable coordinates, perhaps the method of eigenfunction expansions can be used. Traditionally, complex eigenvalues (Sect. 6.2) and related complex (Papkovich-Fadle) eigenfunctions are used. Further, in order to satisfy conditions prescribed at the ends, a "biorthogonality" relation of the eigenfunctions is needed to determine the expansion coefficients.

It would be easier to use real eigenvalues and real eigenfunctions for Stokes flow problems. Also, only orthogonality instead of biorthogonality is required. The following are two examples that illustrate the method. Due to length limitations, the final results are not presented.

6.11.1 Stokes Flow in a Lid-Driven Rectangular Cavity

Figure 6.14 shows a cavity of depth bL and width $2L$. The top surface slides laterally with velocity U. Normalize with L and U. The Stokes equation is Eq. (6.1), with the boundary conditions

$$\psi(x, 0) = 0, \quad \psi(1, y) = 0, \quad \psi(-1, y) = 0, \quad \psi(x, b) = 0 \tag{6.117}$$

$$\psi_y(x, 0) = 1, \quad \psi_x(1, y) = 0, \quad \psi_x(-1, y) = 0, \quad \psi_y(x, b) = 0 \tag{6.118}$$

Fig. 6.14 A rectangular cavity driven by lateral motion of the top plate

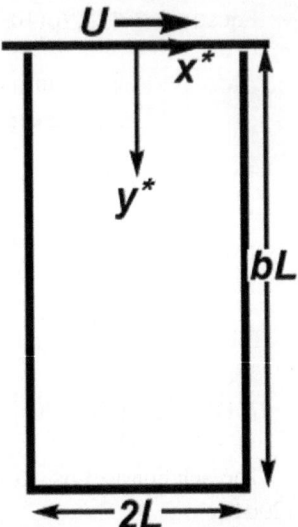

The general solution is given in Eq. (6.2). The solution that satisfies Eq. (6.117) is expressed in Fourier eigenfunctions

$$\psi = \sum_{n=1}^{\infty} \cos(\alpha_n x) \left\{ \begin{array}{c} A_n[b\cosh(\alpha_n b)\sinh(\alpha_n y) - \sinh(\alpha_n b)y\cosh(\alpha_n y)] \\ +B_n(b-y)\sinh(\alpha_n y) \end{array} \right\}$$

$$+ \sum_{m=1}^{\infty} C_m \sin(\beta_m y)[\sinh(\beta_m)\cosh(\beta_m x) - \cosh(\beta_m)x\sinh(\beta_m x)] \qquad (6.119)$$

Here $\alpha_n = \left(n - \frac{1}{2}\right)\pi$, $\beta_m = m\pi/b$, and A_n, B_n, C_m are coefficients to be determined. Equation (6.118a) gives

$$1 = \sum \cos(\alpha_n x)\{A_n[\alpha_n b\cosh(\alpha_n b) - \sinh(\alpha_n b)] + B_n b\alpha_n\}$$

$$+ \sum C_m \beta_m[\sinh(\beta_m)\cosh(\beta_m x) - \cosh(\beta_m)x\sinh(\beta_m x)] \qquad (6.120)$$

Equation (6.118d) gives

$$0 = \sum \cos(\alpha_n x)\{A_n[\alpha_n b - \cosh(\alpha_n b)\sinh(\alpha_n b)] - B_n \sinh(\alpha_n b)\}$$

$$+ \sum C_m \beta_m \cos(\beta_m b)[\sinh(\beta_m)\cosh(\beta_m x) - \cosh(\beta_m)x\sinh(\beta_m x)] \qquad (6.121)$$

Equations (6.118b, 6.118c) give

$$0 = \sum -\alpha_n \sin(\alpha_n) \left\{ \begin{array}{c} A_n[b\cosh(\alpha_n b)\sinh(\alpha_n y) - \sinh(\alpha_n b)y\cosh(\alpha_n y)] \\ + B_n(b-y)\sinh(\alpha_n y) \end{array} \right\}$$

$$+ \sum -C_m \sin(\beta_m y)[\beta_m + \sinh(\beta_m)\cosh(\beta_m)] \qquad (6.122)$$

Multiply Eq. (6.120) by $\cos(\alpha_{\hat{n}}x)$, integrate from 0 to 1, and then let $\hat{n} = n$, resulting in

$$\frac{\sin(\alpha_n)}{\alpha_n} = \frac{1}{2}\{A_n[\alpha_n b\cosh(\alpha_n b) - \sinh(\alpha_n b)] + B_n b\alpha_n\} + \sum C_m \beta_m I_{mn},$$

$$I_{mn} = \frac{-2(-1)^n \alpha_n \beta_m \cosh^2(\beta_m)}{(\alpha_n^2 + \beta_m^2)^2} \qquad (6.123)$$

Similarly, Eq. (6.121) gives

$$0 = \frac{1}{2}\{A_n[\alpha_n b - \cosh(\alpha_n b)\sinh(\alpha_n b)] - B_n \sinh(\alpha_n b)\} + \sum C_m \beta_m (-1)^m I_{mn} \qquad (6.124)$$

Multipling Eq. (6.122) by $\sin(\beta_{\hat{m}} y)$ and integrating from 0 to b, then letting $\hat{m} = m$ give

$$0 = \sum \alpha_n (-1)^n (A_n J_{mn} + B_n K_{mn}) - \frac{b}{2}C_m[\beta_m + \cosh(\beta_m)\sinh(\beta_m)]$$

$$J_{mn} = \frac{-2(-1)^m \alpha_n \beta_m \sinh^2(\alpha_n b)}{(\alpha_n^2 + \beta_m^2)^2}, \quad K_{mn} = \frac{-2\alpha_n \beta_m[-1 + (-1)^m \cosh(\alpha_n b)]}{(\alpha_n^2 + \beta_m^2)^2} \qquad (6.125)$$

Now truncate n to N terms and m to M terms. Equations (6.123, 6.124, 6.125) represent $2N + M$ linear algebraic equations which can be easily inverted for the $2N + M$ unknowns A_n, B_n, C_m.

6.11.2 Flow Through an Array of Strips

Figure 6.15a shows the cross section of a doubly infinite array of solid strips among which fluid is forced through. The arrangement models flow through ordered porous media. The Stokes equation describes the flow since the Reynolds number is very small.

Let the period of the strips transverse to the flow be $2L$, and the longitudinal period be $2aL$. Each strip has width $2bL$ and negligible thickness. A representative cell is enlarged in Fig. 6.15b, where all lengths have been normalized by L.

The governing equation is Eq. (6.1) with the boundary conditions

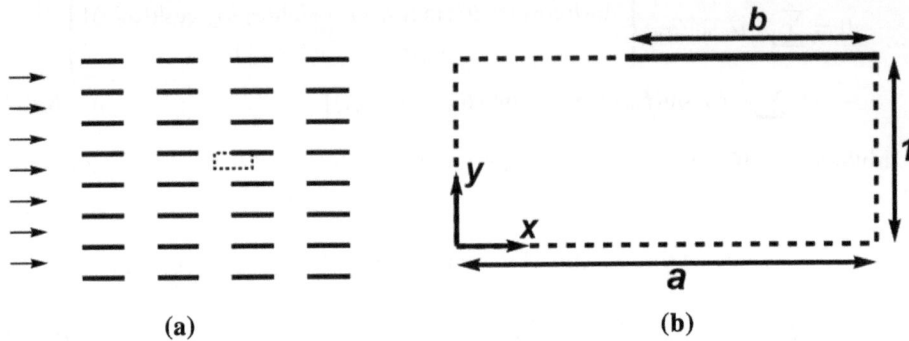

Fig. 6.15 a Cross section of an infinite array of strips **b** A representative cell

$$\psi_x(0, y) = 0, \quad \psi_{xxx}(0, y) = 0, \quad \psi_x(a, y) = 0, \quad \psi_{xxx}(a, y) = 0 \qquad (6.126)$$

$$\psi(x, 0) = 0, \quad \psi_{yy}(x, 0) = 0, \quad \psi(x, 1) = 1, \qquad (6.127)$$

$$\begin{cases} \psi_{yy}(x, 1) = 0, \quad 0 \le x < a - b \\ \psi_y(x, 1) = 0, \quad a - b < x \le a \end{cases} \qquad (6.128)$$

Most of these conditions reflect symmetry or anti-symmetry. The solution satisfying Eqs. (6.126, 6.127) is

$$\psi = y + A_0 y\left(1 - y^2\right) + \sum_{n=1}^{N-1} A_n \cos(\alpha_n x)[\cosh(\alpha_n) \sinh(\alpha_n y) - \sinh(\alpha_n) y \cosh(\alpha_n y)]$$

$$(6.129)$$

where $\alpha_n = n\pi/a$, and A_n are coefficients truncated to N unknowns. Equation (6.128) however is a mixed boundary condition. The easiest way is to use point match as follows. Let

$$x_i = \frac{(i - 0.5)a}{N}, \quad i = 1 \text{ to } N \qquad (6.130)$$

Equations (6.128, 6.129) yield

$$0 = -6A_0 - \sum_{1}^{N-1} 2A_n \cos(\alpha_n x_i)\alpha_n \sinh^2(\alpha_n), \quad i = 1 \text{ to } N, \quad x_i < a - b \qquad (6.131)$$

$$-1 = -2A_0 + \sum_{1}^{N-1} A_n \cos(\alpha_n x_i)[\alpha_n - \sinh(\alpha_n)\cosh(\alpha_n)],$$

$$i = 1 \text{ to } N, \quad x_i > a - b \tag{6.132}$$

There are N linear algebraic equations and N unknowns. Accuracy can be improved by increasing N, since the convergence of Fourier series is well known.

Exercises

(6.1) Consider pulling a flexible sheet around a corner (where one can put a small roller). In Fig. 6.2, let the slanted plate move towards the corner with velocity U and the horizontal plate move away from the corner with the same speed.

(6.2) Consider circular scraping. Envision a diagonal scraper on a rotating disk.

(6.3) Is there a Stokes solution for a single rotating disk analogous to the von Karman solution of Sect. 2.2.5?

(6.4) Solve the problem of uniform Stokes flow over a spherical bubble. Assume the shear stress on the bubble is zero. Is there a drag?

(6.5) Solve the Stokes flow generated by a disk which is torsionally oscillating in its own plane.

(6.6) Using the narrow gap approximation, find the lift and drag when the top plate is described by $h = e^{-\beta x}$.

(6.7) Find the second order solution to the Couette flow of Fig. (6.12). What is the increased drag due to the bottom roughness?

(6.8) Set up the problem of Fig. 6.15a where the mean flow is vertical in the y direction.

Notes

Best references for Stokes flow are Happel and Brenner (1973), Sherman (1990). Complex variable solutions of Stokes flow and lubrication theory are discussed in Langlois and Deville (2014). Boundary perturbations can be found in Wang (1978, 2004). Eigenfunction expansions were illustrated in Wang (1999).

References

J. Happel, H. Brenner, *Low Reynolds Number Hydrodynamics*, 2nd edn. (Kluwer, Dordrecht, Netherlands, 1973)

W.E. Langlois, M.O. Deville, *Slow Viscous Flow*, 2nd edn. (Springer, NY, 2014)

F.S. Sherman, *Viscous Flow* (McGraw-Hill, NY, 1990)

C.Y. Wang, Phys. Fluids **16**, 2136–2139 (2004)

C.Y. Wang, Phys. Fluids **21**, 697–698 (1978)

C.Y. Wang, ZAMP **50**, 982–998 (1999)

Slip Flow

7

The study of fluid flow with boundary slip has become important due to the advances in micro-fluidics. Applications include drag reduction for superhydrophobic microfluidics and moving marine bodies. Surface slip also occurs for rough or chemically treated surfaces, and for particular fluids such as rarefied gases and blood, where a particle-free layer exists near the surface.

In slip flow, the no-slip boundary condition is replaced by Navier's condition (proposed in the nineteenth century when the no-slip condition was not quite established), that the slip velocity is proportional to the local shear stress

$$u_s^* = N\tau_{sn}^* \tag{7.1}$$

The proportionality constant N can be obtained experimentally or theoretically. However, Navier's condition does not apply to unsteady flows.

In what follows, we shall illustrate different analytic solutions of steady slip flow.

7.1 Slip Flow in Ducts

For parallel flow Eq. (7.1) can be replaced by

$$w^* = N\mu \frac{\partial w^*}{\partial \hat{n}^*} \tag{7.2}$$

where w^* is the longitudinal velocity, μ is the fluid viscosity and $\hat{n}^*$ is the outward normal direction. Normalize with a characteristic length L and the velocity by $\frac{GL^2}{\mu}$, where G is the (negative) pressure gradient. The N-S equation reduces to

$$\nabla^2 w = -1 \tag{7.3}$$

© The Author(s), under exclusive license to Springer Nature Switzerland AG 2024
C. Y. Wang, *Essential Analytic Laminar Flow*, Synthesis Lectures on Engineering, Science, and Technology, https://doi.org/10.1007/978-3-031-36449-5_7

with Navier's condition on the wall

$$w + \lambda \frac{\partial w}{\partial \hat{n}} = 0 \qquad (7.4)$$

Here $\lambda = N\mu/L$ is the normalized slip coefficient, or normalized slip length.

For steady flow in long ducts, a non-dimensional measure of the resistance is the Poiseuille number (which supplants the friction factor- Reynolds number product), defined as

$$Po = \frac{|G^*|D_h^{*2}}{2\mu V^*} \qquad (7.5)$$

Here, G^* is the forcing pressure gradient, V^* is the average velocity, $D_h^* = 4A^*/P^*$ is the hydraulic diameter, where A^* and P^* are the area and the perimeter length of the cross section. Equation (7.5) becomes

$$Po = \frac{8A^3}{P^2 Q} \qquad (7.6)$$

where the area and perimeter have been normalized by the length scale and Q is the flow rate normalized by $\frac{|G^*|L^4}{\mu}$.

For slip flow the Poiseuille number is a function of the cross-section geometry, the slip coefficient, but not the length scale.

We shall present two solutions. Some references of slip flow in ducts were given by Wang (2012).

7.1.1 Slip Flow in an Equilateral Triangular Duct

Consider an equilateral triangle with Cartesian vertices at $(-2, 0)$, $(1, \sqrt{3})$, $(1, -\sqrt{3})$. The length scale is the radius of the inscribing circle. The solution (Wang 2012) is

$$w = -\frac{1}{4}(x^2 + y^2) + c_1 + c_2(x^3 - 3xy^2)$$

$$c_1 = \frac{2 + 6\lambda + 3\lambda^2}{6(1 + \lambda)}, c_2 = \frac{-1}{12(1 + \lambda)} \qquad (7.7)$$

The Poiseuille number is

$$Po = \frac{40(1 + \lambda)}{(3 + 15\lambda + 10\lambda^2)} \qquad (7.8)$$

The Poiseuille number decreases as boundary slip is increased.

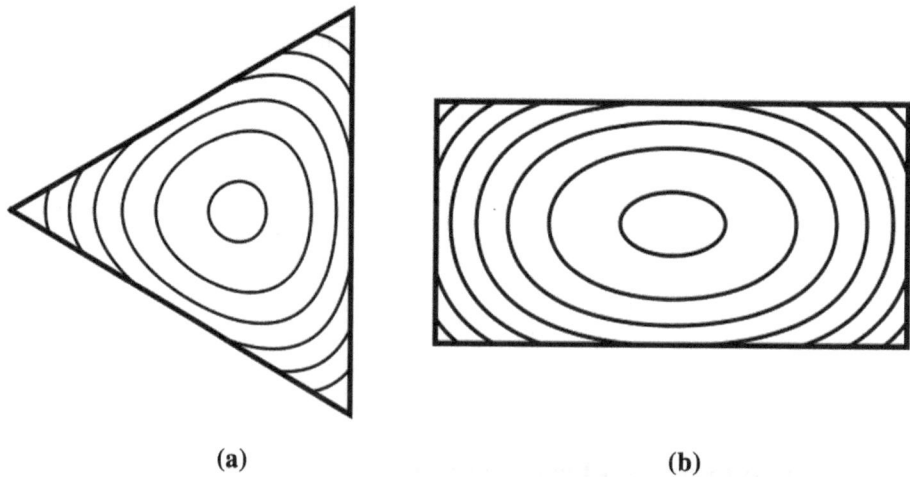

(a)   **(b)**

Fig. 7.1 Constant velocity curves for $\lambda = 1$ **a** Equilateral triangular duct **b** 2:1 rectangular duct

Figure 7.1a shows the constant velocity curves for slip flow in an equilateral triangular duct. Notice the velocity is not zero on the wall as in the no-slip case. The minimum velocity, which is not zero, occurs at the corners.

7.1.2   Slip Flow in a Rectangular Duct

As in the no-slip case, the solution to the flow in a rectangular duct can only be in series form (Ebert and Sparrow 1965). Consider a rectangular duct of width $2L$ and height $2bL$. Place Cartesian coordinates at the centroid. Let the normalized velocity be

$$w = \sum_{n=1}^{\infty} A_n \cos(\alpha_n x) f_n(y) \tag{7.9}$$

Equation (7.4) at $x = 1$ gives the characteristic equation

$$\cos(\alpha_n) - \lambda \alpha_n \sin(\alpha_n) = 0 \tag{7.10}$$

The eigenvalues α_n can be obtained by root search or asymptotically for larger n. One can also show orthogonality

$$\int_0^1 \cos(\alpha_n x) \cos(\alpha_m x) dx = \begin{cases} \frac{1}{2} + \frac{\sin(\alpha_n)\cos(\alpha_n)}{2\alpha_n}, & m = n \\ 0, & m \neq n \end{cases} \tag{7.11}$$

Using Eq. (7.11), unity is expanded and the coefficients A_n are determined

$$1 = \sum_{n=1}^{\infty} A_n \cos(\alpha_n x), \ A_n = \frac{2\sin(\alpha_n)}{\alpha_n + \sin(\alpha_n)\cos(\alpha_n)} \tag{7.12}$$

Then, Eqs. (7.3), (7.4), (7.9), (7.12) yield

$$f_n'' - \alpha_n^2 f_n = -1, \quad f_n(b) + \lambda f_n\prime(b) = 0 \tag{7.13}$$

The solution is

$$f_n = \frac{1}{\alpha_n^2}\left[1 - \frac{\cosh(\alpha_n y)}{\cosh(\alpha_n b) + \lambda\alpha_n \sinh(\alpha_n b)}\right] \tag{7.14}$$

Equation (7.9), plotted in Fig. 7.1b, shows the velocity distribution.

7.2   Some Exact Slip Flow Solutions

Due to the slip boundary conditions, exact solutions for (non- parallel) slip flow are much fewer than their no-slip counter parts. There are stretching flows with slip, stagnation flows with slip, rotating disks with slip and slip in a rotating channel. We shall mention some typical solutions.

7.2.1   Stretching Sheet with Slip and Stagnation Flow with Slip

The two-dimensional no-slip stretching sheet was introduced in Sect. 2.2.1. Using a normalized stream function $\psi = xf(y)$ the N-S equation gives

$$f'' - ff'' = f''' \tag{7.15}$$

The slip boundary conditions are

$$f'(0) - 1 = \lambda f''(0), \ f(0) = 0, \ f'(\infty) = 0 \tag{7.16}$$

Notice the subtraction of unity is due to the velocity difference in Navier's condition. Andersson (2002) found the exact solution

$$f = \beta(1 - e^{-\beta y}) \tag{7.17}$$

where β satisfies the cubic

$$\lambda\beta^3 + \beta^2 - 1 = 0 \tag{7.18}$$

Stagnation flow with slip is also governed by Eq. (7.15). The boundary conditions are

$$f'(0) = \lambda f''(0), \, f(0) = 0, \, f'(\infty) = 1 \tag{7.19}$$

There are no closed-form solutions, and the similarity equation was integrated by Wang (2003a, b, c).

7.2.2   Slip on a Rotating Disk

This problem is an extension of Von Karman's exact solution (Sect. 2.2.5). The governing equations are Eqs. (2.32) but with the slip boundary conditions

$$f'(0) = \lambda_1 f''(0), \, g(0) - 1 = \lambda_2 g'(0), \, f(0) = 0, \, f'(\infty) = 0, \, g(\infty) = 0 \tag{7.20}$$

where λ_1 is the slip coefficient in the radial direction and λ_2 is that in the azimuthal direction (anisotropic as if on a phonograph record). Miklavcic and Wang (2004) integrated the similarity equations.

7.2.3   Slip Flow in a Rotating Channel

The pressure driven flow between parallel plates is the well-known Poiseuille flow. The whole system may be rotated about a normal axis as in a centrifuge (Wang 2013). Normalize lengths by the distance H between the plates, the velocities by H^2(pressure gradient)/(viscosity) and the rotation rate by (kinematic viscosity)/H^2. The governing equations are

$$1 + \frac{d^2u}{dz^2} = -2\Omega v, \quad \frac{d^2v}{dz^2} = 2\Omega u \tag{7.21}$$

where (u, v, w) are velocities in the Cartesian (x, y, z) directions respectively, and Ω is the normalized angular velocity. The slip boundary conditions on the plates at $z = 0, 1$ are

$$u(0) = \lambda \frac{du}{dz}(0), \, u(1) = -\lambda \frac{du}{dz}(1), \, v(0) = \lambda \frac{dv}{dz}(0), \, v(1) = -\lambda \frac{dv}{dz}(1) \tag{7.22}$$

Construct a complex velocity by

$$\phi = u + iv \tag{7.23}$$

where $i = \sqrt{-1}$. Then Eqs. (7.21, 7.22) become

$$\frac{d^2\phi}{dz^2} - 2\Omega i\phi = -1, \quad \phi(0) = \lambda \frac{d\phi}{dz}(0), \quad \phi(1) = -\lambda \frac{d\phi}{dz}(1) \tag{7.24}$$

The solution is

$$\phi = \frac{-i}{2\Omega}\left[1 - \frac{e^{s(1-z)} + e^{sz}}{1 - \lambda s + (1 + \lambda s)e^s}\right], \quad s = (1 + i)\sqrt{\Omega} \tag{7.25}$$

Notice that u is the velocity in the direction of the pressure gradient and v is the secondary transverse flow due to Coriolis force.

7.3   Stokes Slip Flow

These slip flows are not parallel or exact N-S solutions. The nonlinear terms are unimportant due to low Reynolds numbers, which are reasonable in microfluidics. In this section we shall illustrate with some examples of treating Stokes slip flow.

7.3.1   Stokes Slip Flow over a Sphere

This problem was originally solved by Stokes, again when the no-slip condition was not in vogue. The Stokes equation in spherical coordinates is Eq. (6.6). Similar to Eq. (6.27), the bounded solution which gives uniform velocity at infinity is

$$\psi = \frac{1}{2}\sin^2(\theta)\left(\varrho^2 + \frac{c_1}{\varrho} + c_2\varrho\right) \tag{7.26}$$

where we have normalized with infinity velocity U and sphere radius a. The shear stress Eq. (1.51) is

$$\tau_{\varrho\theta} = 2d_{\varrho\theta} = \frac{1}{\varrho}\frac{\partial u}{\partial\theta} + \frac{\partial v}{\partial\varrho} - \frac{v}{\varrho} \tag{7.27}$$

where

$$u = \frac{1}{\varrho^2\sin(\theta)}\psi_\theta, \quad v = \frac{-1}{\varrho\sin(\theta)}\psi_\varrho \tag{7.28}$$

Since $u = 0$ on the sphere, Navier's condition is (the outward normal is into the sphere)

$$v = \lambda\left(\frac{\partial v}{\partial\varrho} - \frac{v}{\varrho}\right) \tag{7.29}$$

The boundary conditions are that on $\varrho = 1$

$$\psi_r = 0, \quad \psi_\theta = \lambda\left(\psi_{\varrho\varrho} - 2\psi_\varrho\right) \tag{7.30}$$

Equations (7.26, 7.30) yield

$$c_1 = \frac{1}{2(1+3\lambda)}, \quad c_2 = \frac{-3(1+2\lambda)}{2(1+3\lambda)} \tag{7.31}$$

The drag reduction can be found similar to the method in Sect. 6.6.

7.3.2 Slip Flow in a Wavy Channel

For the flow in a duct with wavy or striated walls, one can use perturbation to simplify the boundary conditions as in Sect. 6.10. For slip flow, if the striations are parallel to the flow, the surface stress in Navier's condition can be supplanted by a normal derivative Eq. (7.4). However, if the striations are transverse to the flow, the stress tensor also contains other terms. Consider a two-dimensional surface in Cartesian

$$g(x, y) = y - 1 - \varepsilon f(x) = 0 \tag{7.32}$$

The normalized Navier condition is

$$u_s = \lambda \tau_{ns} \tag{7.33}$$

where (n, s) are intrinsic coordinates normal and tangential to the surface. The velocity u_s, being a vector, can be decomposed into the Cartesian directions as follows.

$$u_s = \frac{\partial \psi}{\partial \hat{n}} = \hat{n} \cdot \nabla \psi = \frac{\nabla g}{|\nabla g|} \cdot \nabla \psi = \frac{\psi_y - \varepsilon f' \psi_x}{\sqrt{1 + (\varepsilon f')^2}} \tag{7.34}$$

If the stream function is expanded

$$\psi = \psi_0 + \varepsilon \psi_1 + \varepsilon^2 \psi_2 + \cdots \tag{7.35}$$

then

$$u_s = \psi_{0y} + \varepsilon \left(\psi_{1y} - f' \psi_{0x} \right) + \varepsilon^2 \left(\psi_{2y} - f' \psi_{1x} - \frac{1}{2} f'^2 \psi_{0y} \right) + \cdots \tag{7.36}$$

But the stress τ_{ns} is a tensor and follows tensor rotation rules. Wang (2011) found

$$\begin{aligned} \tau_{ns} = &\psi_{yy} - \psi_{xx} - \frac{4\varepsilon f' \psi_{xy}}{\left[1 - \varepsilon^2 (f')^2\right]} = \psi_{0yy} - \psi_{0xx} \\ &+ \varepsilon \left(\psi_{1yy} - \psi_{1xx} - 4 f' \psi_{0xy} \right) \\ &+ \varepsilon^2 \left(\psi_{2yy} - \psi_{2xx} - 4 f' \psi_{1xy} \right) + \cdots \end{aligned} \tag{7.37}$$

Figure 7.2 shows the transverse flow through a wavy channel. The perturbed channel walls are at

Fig. 7.2 Transverse flow in a wavy channel (bottom corrugations are out of phase)

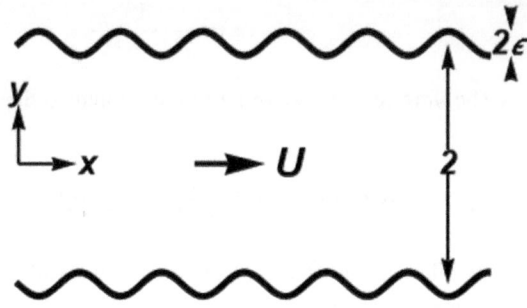

$$y = 1 + \varepsilon f_1(x), \quad f_1 = \sin(\alpha x)$$

$$y = -1 + \varepsilon f_2(x), \quad f_2 = \pm \sin(\alpha x) \tag{7.38}$$

where the top sign is used when the corrugations are in phase and the bottom sign when they are out of phase.

For Stokes flow, the governing equation is Eq. (6.100) with the normalized slip boundary conditions

$$\psi = 1, u_s = \lambda \tau_{ns} \text{ on } y = 1 + \varepsilon \sin(\alpha x) \tag{7.39}$$

$$\psi = -1, u_s = -\lambda \tau_{ns} \text{ on } y = -1 \pm \varepsilon \sin(\alpha x) \tag{7.40}$$

As in Eq. 6.103, Taylor series is used to move the boundary conditions to $y = \pm 1$. Then each order of ε is solved successively by perturbation (Appendix B).

It is found that the zeroth order is slip flow in a smooth channel

$$\psi_0 = 2y - y^3 - \left(\frac{1 + 6\lambda}{2 + 6\lambda} \right)(y - y^3) \tag{7.41}$$

The first order ψ_1, proportional to $\sin(\alpha x)$, gives the correction to shear stress. The second order, proportional to $\sin^2(\alpha x)$ having a non-periodic component, gives the changes in the resistance or the mean pressure gradient.

7.3.3 Stokes Slip Flow Through a Filter

Figure 7.3a shows the cross section of a filter which consists of a column of parallel slats (like a Venetian blind). Normalize with the free stream velocity U and the half distance between two slats L. Instead of using point match as in Sect. 6.11.2, we shall use domain decomposition (Wang 2010).

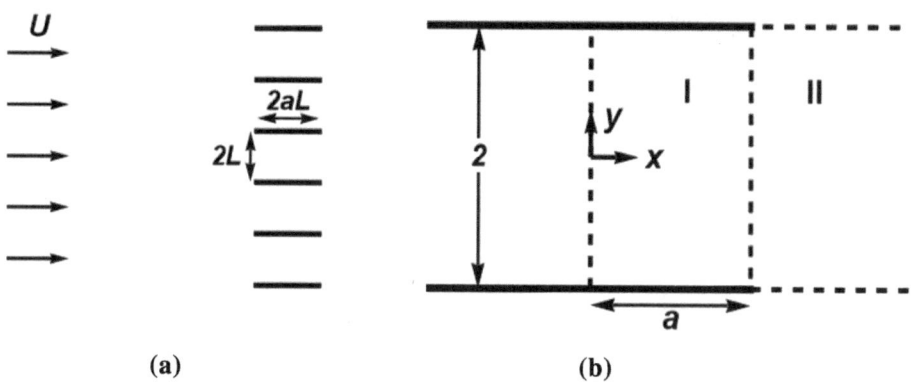

Fig. 7.3 **a** Fluid forced through a column of slats with surface slip **b** Decomposition into two regions

Figure 7.3b shows a repeated cell divided into two regions. Region I includes half of the area between the slats and Region II is a semi-infinite fluid domain. The normalized stream functions ψ_I, ψ_{II} satisfy the Stokes biharmonic equation, Eq. (6.1). For Region I the boundary conditions are

$$\psi_I(x, \pm 1) = \pm 1 \tag{7.42}$$

$$\psi_{Iy}(x, \pm 1) = \mp \lambda \psi_{Iyy}(x, \pm 1) \tag{7.43}$$

The general solution, even in x and odd in y, that satisfies Eq. (7.42) is

$$
\begin{aligned}
\psi_I(x, y) = {} & \psi_0(y) + \sum_{n=1}^{N} \sin(\alpha_n y)[A_n \cosh(\alpha_n x) + B_n x \sinh(\alpha_n x)] + \\
& C_0(y - y^3) + \sum_{m=1}^{M-1} C_m \cos(\beta_m x)[\coth(\beta_m) \sinh(\beta_m y) - y \cosh(\beta_m y)]
\end{aligned}
\tag{7.44}
$$

Here $\alpha_n = n\pi$, $\beta_m = m\pi/a$ and

$$\psi_0 = \frac{3(1 + 2\lambda)y - y^3}{2(1 + 3\lambda)} \tag{7.45}$$

is the stream function solution for slip flow between infinite parallel plates. Substitute Eq. (7.44) into Eq. (7.43) and equate each Fourier component in x

$$\int_0^a \psi_{Iy}(x, 1)\cos(\beta_m x)dx = -\lambda \int_0^a \psi_{Iyy}(x, 1)\cos(\beta_m x)dx \quad m = 0 \text{ to } M - 1 \tag{7.46}$$

For Region II, the boundary conditions are

$$\psi_{II}(x, \pm 1) = \pm 1 \tag{7.47}$$

$$\psi_{IIyy}(x, \pm 1) = 0 \tag{7.48}$$

The general solution that satisfies Eqs. (7.47), (7.48) is

$$\psi_{II}(x, y) = y + \sum_{n=1}^{N} \sin(\alpha_n y)[D_n + E_n(x - a)]e^{-\alpha_n(x-a)} \tag{7.49}$$

where the y term denotes uniform flow to which the stream function approaches at infinity. At the interface of the two Regions, the stream function is continuous and the Fourier components in y are equated

$$\int_0^1 [\psi_I(a, y) - \psi_{II}(a, y)] \sin(\alpha_n y)dy = 0, \quad n = 1 \text{ to } N \tag{7.50}$$

$$\int_0^1 [\psi_{Ix}(a, y) - \psi_{IIx}(a, y)] \sin(\alpha_n y)dy = 0, \quad n = 1 \text{ to } N \tag{7.51}$$

$$\int_0^1 [\psi_{Ixx}(a, y) - \psi_{IIxx}(a, y)] \sin(\alpha_n y)dy = 0, \quad n = 1 \text{ to } N \tag{7.52}$$

$$\int_0^1 [\psi_{Ixxx}(a, y) - \psi_{IIxxx}(a, y)] \sin(\alpha_n y)dy = 0, \quad n = 1 \text{ to } N \tag{7.53}$$

Equations (7.46, 7.50–7.53) are $M + 4N$ linear algebraic equations for the $M + 4N$ unknown coefficients C_m, A_n, B_n, D_n, E_n, and can be easily solved.

After the stream function is found, the pressure, shear, and thus the resistance of the filter can be determined.

7.4   Slip Flow in Curved Ducts

Curved ducts are essential for redirecting the flow. In this section we consider ducts whose centerline is curved while the cross section is invariant. The method is to use perturbation about the straight duct solution.

7.4.1   Stokes Slip Flow in a Torus

Consider the curved duct with a circular cross section (a torus) whose geometry is shown in Fig. 1.2. Using Dean's orthogonal coordinates Eq. (1.58), we find

$$ds^{*2} = dr^{*2} + r^{*2}d\phi^2 + \left[R + r^*\cos(\theta)\right]^2 d\theta^2 \tag{7.54}$$

which leads to the scale factors

$$h_{(r)} = 1, h_{(\phi)} = r^*, h_{(\theta)} = \left[R + r^*\cos(\theta)\right] \equiv H^* \tag{7.55}$$

From Eq. (1.15), the Stokes equation is

$$-\frac{1}{\mu}\nabla p^* = -\nabla^2 u^* = \nabla \times \left(\nabla \times u^*\right) \tag{7.56}$$

For Stokes flow in a duct with constant curvature, the velocity $u^* = (0, 0, w^*)$ only has the azimuthal component. Using the scale factors (Sect. 1.2), Eq. (7.56) becomes

$$-\frac{1}{\mu H^*}p_\theta^* = -\frac{1}{r^*}\left[\frac{r^*}{H^*}(H^*w^*)_{,r^*}\right]_{r^*} - \frac{1}{r^*}\left[\frac{1}{r^*H^*}(H^*w^*)_\phi\right]_\phi \tag{7.57}$$

Now normalize lengths by the radius of curvature R and the velocity by $\frac{GR^2}{\mu}$, where G is the pressure gradient along the centerline, and drop primes. For small cross section radius a, the tube radius should be normalized by a instead of R,

$$\frac{r^*}{R} = \frac{a}{R}\frac{r^*}{a} \equiv \varepsilon\eta, \quad \varepsilon = \frac{a}{R} \ll 1 \tag{7.58}$$

Then

$$H = \frac{H^*}{R} = 1 + \varepsilon\eta\cos(\phi) \tag{7.59}$$

We also set the normalized pressure gradient to be -1, and due to the small radius

$$\frac{w^*}{GR^2/\mu} = \varepsilon^2\frac{w^*}{Ga^2/\mu} = \varepsilon^2 w \tag{7.60}$$

Equation (7.57) becomes

$$\frac{1}{H} = -\frac{1}{\eta}\left[\frac{\eta}{H}(Hw)_\eta\right]_\eta - \frac{1}{\eta}\left[\frac{1}{\eta H}(Hw)_\phi\right]_\phi \tag{7.61}$$

Using Eq. (1.31), Navier's condition is, on the surface at $\eta = 1$,

$$w = -\lambda\tau_{r\theta} = -\lambda H\left(\frac{w}{H}\right)_\eta, \quad \lambda = \frac{N\mu}{a} \tag{7.62}$$

Now expand

$$w = w_0 + \varepsilon w_1 + \varepsilon^2 w_2 + \cdots \tag{7.63}$$

When $\varepsilon = 0$ (straight tube), w_0 is a function of radius η only. The zeroth order of Eq. (7.61) is

$$1 = -\frac{1}{\eta}(\eta w_{0\eta})_\eta \tag{7.64}$$

The boundary condition from Eq. (7.62) is

$$w_0(1) = -\lambda w_{0\eta}(1) \tag{7.65}$$

The solution is the slip flow in a straight tube

$$w_0 = \frac{1 + 2\lambda - \eta^2}{4} \tag{7.66}$$

The first order equation is

$$\frac{1}{\eta}(\eta w_{1\eta})_\eta + \frac{1}{\eta^2}w_{1\phi\phi} = \cos(\phi)(\eta - w_{0\eta}) \tag{7.67}$$

The boundary condition is

$$w_1(1, \phi) = -\lambda[w_{1\eta}(1, \phi) - \cos(\phi)w_0(1)] \tag{7.68}$$

The solution is

$$w_1 = \cos(\phi)F(\eta) \tag{7.69}$$

where F is a polynomial in η. Consult a similar procedure given in Wang (2003c).

7.4.2 Slip Flow in a Channel Bend

Figure 7.4 shows a curved channel where all lengths have been normalized by the half constant width. Let (s, n) be coordinates along and normal to the centerline whose variable curvature is $K(s)$. The scale factors are $H \equiv (1 + Kn)$ and 1. If (u, v) are the corresponding velocity components, Sect. 1.2 shows the continuity equation is satisfied by the stream function $\psi(s, n)$

$$(u, v) = \left(\psi_n, \frac{-1}{H}\psi_s\right) \tag{7.70}$$

Fig. 7.4 A channel bend with varying curvature K

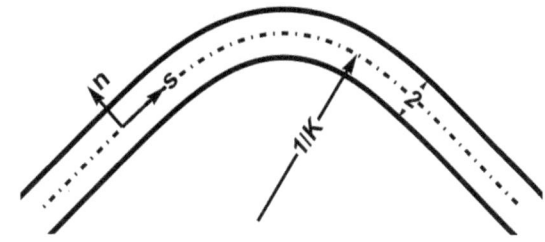

The Stokes equation is

$$\nabla^4 \psi = 0, \quad \nabla^2 = \frac{1}{H}\left[\frac{\partial}{\partial s}\left(\frac{1}{H}\frac{\partial}{\partial s}\right) + \frac{\partial}{\partial n}\left(H\frac{\partial}{\partial n}\right)\right] \tag{7.71}$$

The boundary conditions are

$$\psi(s, \pm 1) = 0 \tag{7.72}$$

$$\psi_n(s, \pm 1) = \mp \lambda H\left(\frac{1}{H}\psi_n\right)_n\bigg|_{n=\pm 1} \tag{7.73}$$

Next assume curvature is small and slowly varying (Van Dyke 1983), i.e.,

$$K = \varepsilon k(t), \quad t = \varepsilon s \tag{7.74}$$

Also the stream function is expanded about that of the straight channel

$$\psi = \psi_0(n) + \varepsilon \psi_1(t, n) + \varepsilon^2 \psi_2(t, n) + \cdots \tag{7.75}$$

The zeroth orders of Eqs. (7.71–7.73) give

$$\psi_0'''' = 0, \quad \psi_0(\pm 1) = \pm 1, \quad \psi_0'(\pm 1) \pm \lambda \psi_0''(\pm 1) = 0 \tag{7.76}$$

The solution is the slip flow in a straight channel

$$\psi_0 = \frac{n(3 + 6\lambda - n^2)}{2(1 + 3\lambda)} \tag{7.77}$$

The first order is determined by

$$\psi_{1nnmn} + 2k\psi_0''' = 0, \quad \psi_1(t, \pm 1) = 0, \quad \psi_{1n}(t, \pm 1) \pm \lambda \psi_{1nn}(t, \pm 1) = \pm \lambda k \psi_0'(\pm 1) \tag{7.78}$$

The solution is

$$\psi_1 = k(t) \frac{(1-n^2)[1 + 5\lambda - 6\lambda^2 - (1+\lambda)n^2]}{4(1+\lambda)(1+3\lambda)} \tag{7.79}$$

We shall not go into the higher order corrections which have been found by Wang (2020).

Exercises

(7.1) Consider the shear flow over a surface which is covered with a thin layer of a fluid of lower viscosity. Obtain the proportionality constant of Eq. (7.1).

(7.2) Using Navier's condition Eq. (7.4), solve for the slip flow through a circular cylinder. What is the Poiseuille number?

(7.3) Solve the flow due to a plate with anisotropic surface slip which is moving in a rotating system.

(7.4) The Stokes flow over a sphere was studied in Sect. 6.6. Solve for the drag when there is surface slip.

(7.5) Consider the slip flow in a wavy channel parallel to the corrugations, i.e., the flow is in the z direction in Fig. 7.2. What is the resistance or pressure gradient?

Notes

Background of slip flow can be found in Nguyen and Wereley (2006), Rothstein (2010). The slip coefficient for a grooved surface was determined by Wang (2003a). References of slip flow, including the inapplicability of Navier's condition for unsteady slip flow, was given in Ebert and Sparrow (1965), Wang (2012). References for exact slip flow solutions are Andersson (2002), Wang (2003b), Miklavcic and Wang (2004), Wang (2013). Navier's condition on a curved boundary was discussed in Wang (2011). Domain decomposition method was illustrated in Wang (2010). References for slip flow in bends are Wang (2003c, 2020), Van Dyke (1983).

References

H.I. Andersson, Acta Mech. **158**, 121–125 (2002)

W.A. Ebert, E.M. Sparrow, J. Basic Eng. **87**, 1018–1024 (1965)

M. Miklavcic, C.Y. Wang, ZAMP **55**, 235–246 (2004)

N.T. Nguyen, S.T. Wereley, *Fundamentals and Applications of Microfluidics*, 2nd Ed (Artech House, MA, 2006)

J. Rothstein, Ann. Rev. Fluid Mech. **42**, 89–109 (2010)

M. Van Dyke, SIAM, J. Appl. Math. **43**, 696–702 (1983)

C.Y. Wang, Can. J. Chem. Eng. **88**, 335-339 (2010)

C.Y. Wang, Chem. Eng. Comm. **200**, 587–594 (2013)

C.Y. Wang, J. Fluids Eng. **125**, 443–446 (2003c)

C.Y. Wang, J. Fluids Eng. **134**, 094501 (2012)

C.Y. Wang, J. Fluids Eng. **142**, 014504 (2020)
C.Y. Wang, Mech. Res. Comm. **38**, 249–254 (2011)
C.Y. Wang, Phys. Fluids **15**, 1114–1121 (2003a)
C.Y. Wang, ZAMP **54**, 184–189 (2003b)

Darcy–Brinkman Flow

8

Modelling fluid flow through porous media is important in ground water movement, insulation, oil and gas recovery, catalytic converters, filtering etc. The Darcy equation, which assumes the pressure gradient is proportional to the seepage velocity, leads to potential flow, which is discussed in Appendix C. The Darcy equation may not be adequate in describing the flow since the resistance of possible solid boundaries is ignored. Brinkman added a viscous term to the Darcy equation, which is the well-accepted Darcy-Brinkman equation (D–B equation). The assumptions are.

- The porous matrix is saturated with a single Newtonian fluid.
- The velocity is low, such that the nonlinear momentum terms and the Forchheimer drag are negligible.
- The porous matrix is firmly attached to the embedded solid boundary.

The last point is important since freely flowing particles would offer no resistance. Even if the particles are stationary, the matrix should be firmly attached to any solid boundary, otherwise a clear fluid channel occurs along the solid surface and would invalidate the D–B equation (Nield and Bejan 2017). Thus, D–B flow cannot have any relative motion, normal or tangential, between a solid boundary and a porous medium.

8.1 The D–B Equation

The unsteady D–B equation is

$$\frac{\partial \boldsymbol{u}^*}{\partial t^*} + \nabla p^* = \mu_e \nabla^2 \boldsymbol{u}^* - \frac{\mu}{K} \boldsymbol{u}^* \tag{8.1}$$

© The Author(s), under exclusive license to Springer Nature Switzerland AG 2024
C. Y. Wang, *Essential Analytic Laminar Flow*, Synthesis Lectures on Engineering,
Science, and Technology, https://doi.org/10.1007/978-3-031-36449-5_8

Here, the asterisked variables are local averaged values, p is the pressure, $\boldsymbol{u}$ is the velocity vector, t is the time, μ is the fluid viscosity, K is the permeability and μ_e the effective viscosity of the porous matrix. Normalize the time by L^2/μ_e, all lengths by a characteristic length L, the negative pressure gradient by its maximum G_0, the velocity by $\frac{L^2 G_0}{\mu_e}$ and drop primes. Equation (8.1) becomes

$$\nabla^2 \boldsymbol{u} - k^2 \boldsymbol{u} = -\boldsymbol{G} + \boldsymbol{u}_t \tag{8.2}$$

where

$$k = L\sqrt{\mu/\mu_e K} \tag{8.3}$$

is the porous medium factor proportional to the inverse square root of the Darcy number. When $k = 0$, the porous medium is absent, and we recover the unsteady Stokes equation.

For parallel flow Eq. (8.2) gives

$$\nabla^2 w - k^2 w = -G(t) + w_t \tag{8.4}$$

where the ∇^2 operates in the plane normal to the longitudinal velocity w.

For two-dimensional transverse flow, one can define a stream function that satisfies continuity

$$\nabla \cdot \boldsymbol{u} = 0 \tag{8.5}$$

Eliminate pressure and express the D–B equation in terms of stream function (similar to Sect. 1.1).

The D–B equation is

$$L^4 \psi - k^2 L^2 \psi = L^2 \psi_t \tag{8.6}$$

where the operator L^2 depends on the coordinate system used. The solution to Eq. (8.6) is the linear combination of the solutions of

$$L^2 \psi = 0 \quad \text{and} \quad L^2 \psi - k^2 \psi = \psi_t \tag{8.7}$$

For Cartesian coordinates,

$$L^2 = \frac{\partial^2}{\partial x^2} + \frac{\partial^2}{\partial y^2} \tag{8.8}$$

For two dimensional cylindrical coordinates in (r, θ),

$$L^2 = \frac{\partial^2}{\partial r^2} + \frac{1}{r}\frac{\partial}{\partial r} + \frac{1}{r^2}\frac{\partial^2}{\partial \theta^2} \tag{8.9}$$

For axisymmetric cylindrical coordinates in (r, z),

$$L^2 = \frac{\partial^2}{\partial r^2} - \frac{1}{r}\frac{\partial}{\partial r} + \frac{\partial^2}{\partial z^2} \tag{8.10}$$

For axisymmetric spherical coordinates in (ϱ, θ), let $\xi = \cos(\theta)$

$$L^2 = \frac{\partial^2}{\partial \varrho^2} + \frac{1 - \xi^2}{\varrho^2}\frac{\partial^2}{\partial \xi^2} \tag{8.11}$$

We shall show the different solutions to D–B flow in the following sections.

8.2   Closed-Form Solutions of the D–B Equation

Since closed-form solutions have no error, they are important as accuracy standards for numerical or asymptotic solutions. The steady closed-form D–B solutions were reviewed by Wang (2023). Here we shall present some unsteady solutions.

From Eq. (8.7) we find closed-form solutions can only be exponential in time. Let

$$\psi = e^{\lambda t}\overline{\psi}(x) \tag{8.12}$$

where the exponent λ may be real or complex, x are space variables, and only the real part of the product in Eq. (8.12) is relevant.

8.2.1   Parallel Flow

For parallel flow, Let the pressure gradient and the longitudinal velocity be exponential in time

$$G = e^{\lambda t}, \quad w = e^{\lambda t}\overline{w}(x) \tag{8.13}$$

Equation (8.4) then yields the following solutions.

8.2.1.1   D–B Flow Tangent to a Plate
Consider first a porous medium bounded by a single plate. Equations (8.4, 8.13) give

$$\frac{d^2\overline{w}}{dy^2} - k^2\overline{w} = -1 + \lambda\overline{w} \tag{8.14}$$

The solution satisfying the no-slip boundary condition at $y = 0$ and bounded at infinity is

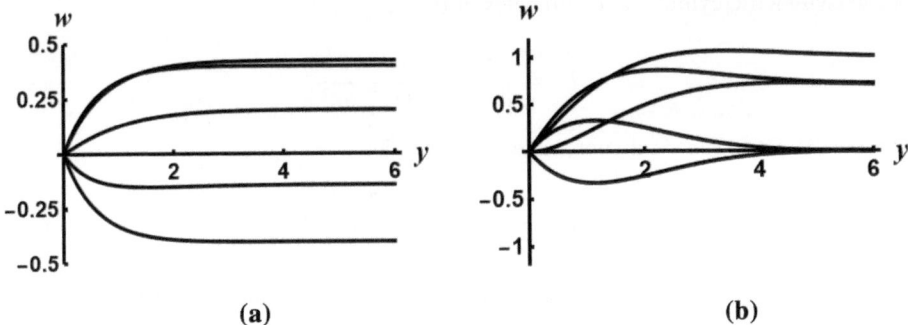

Fig. 8.1 D–B velocity profiles for oscillatory pressure gradient over a plate for $\omega = 1$ **a** $k^2 = 2$, from top: $\frac{t}{\pi} = 0, 0.25, 0.5, 0.75, 1$ **b** $k^2 = 0.02$, from top: $\frac{t}{\pi} = 0.5, 0.25, 0.75, 0, 1$.

$$\overline{w} = \frac{1}{s^2}\left[1 - e^{-sy}\right] \tag{8.15}$$

where

$$s \equiv \sqrt{k^2 + \lambda} \tag{8.16}$$

If λ is real and $\lambda > -k^2$, Eqs. (8.13, 8.15) show the velocity approaches $\frac{e^{\lambda t}}{(k^2+\lambda)}$ for large y. It is an accelerating flow (and an accelerating pressure) when $\lambda > 0$, decelerating flow when $0 > \lambda > -k^2$, and steady flow when $\lambda = 0$. No solution exists when $\lambda \le -k^2$.

If λ is complex, say $\lambda = \omega i$, where ω is the frequency of pressure oscillation and $i = \sqrt{-1}$, a bounded solution is still possible if the real part of $s \equiv \sqrt{k^2 + \lambda}$ is less than zero. Thus,

$$w = \text{Re}\left[e^{i\omega t} \frac{1}{\left(k^2 + i\omega\right)}\left(1 - e^{-\sqrt{k^2+i\omega}\,y}\right)\right] \tag{8.17}$$

Figure 8.1 shows some typical velocity profiles. When the porous medium factor k is large (small permeability), the velocity is in-phase with the pressure. When k is small, the velocity is off phase and overshoots.

8.2.1.2 D–B Flow in a Channel

The solution for unsteady flow in a channel ($|y| \le 1$) filled with a porous medium is

$$\overline{w} = \frac{1}{s^2}\left[1 - \frac{\cosh(sy)}{\cosh(s)}\right] \tag{8.18}$$

The properties are similar to those of the single plate problem, except a solution exists when $s^2 = k^2 + \lambda = 0$. It is the decaying ($\lambda < 0$) Poiseuille flow

$$w = e^{\lambda t} \frac{(1 - y^2)}{2} \tag{8.19}$$

8.2.1.3 D–B Parallel Flow Inside and Outside a Circular Duct

The solutions are

$$\overline{w} = \frac{1}{s^2} \left[1 - \frac{I_0(sr)}{I_0(s)} \right] \tag{8.20}$$

$$\overline{w} = \frac{1}{s^2} \left[1 - \frac{K_0(sr)}{K_0(s)} \right] \tag{8.21}$$

where I_0 and K_0 are modified Bessel functions.

8.2.2 Unsteady Flow in a Porous Rotating Channel

Let the channel walls be at $z = \pm 1$ and rotate as a system about the z axis. Similar to Sect. 2.1.3 the governing equations are

$$u_t - G(t) = u_{zz} - k^2 u + 2\beta v \tag{8.22}$$

$$v_t = v_{zz} - k^2 v - 2\beta u \tag{8.23}$$

Here (u, v) are velocity components in the (x, y) directions, G is the negative pressure gradient in the x direction, and $\beta = \rho L^2 \Omega / \mu_e$ is the nondimensionalization of the rotation rate Ω. Let $G = e^{\lambda t}$ and

$$\phi = u + iv = e^{\lambda t} \overline{\phi}(z) \tag{8.24}$$

Then

$$\lambda \overline{\phi} - 1 = \overline{\phi}_{zz} - k^2 \overline{\phi} - 2i\beta \overline{\phi} \tag{8.25}$$

The solution is

$$\overline{\phi} = \frac{1}{\Lambda^2} \left[1 - \frac{\cosh(\Lambda z)}{\cosh(\Lambda)} \right], \quad \Lambda = \sqrt{k^2 + \lambda + 2i\beta} \tag{8.26}$$

8.2.3 D–B Stagnation Flows

For the stagnation flow on a plate, let the tangential velocity be proportional to x as in Sect. 2.2.2 or

$$\psi = xe^{\lambda t} f(y) \tag{8.27}$$

Equation (8.7) yields the solutions.

$$f = 1, y \text{ and } f = e^{\pm sy} \tag{8.28}$$

The solution that satisfies no-slip and bounded at infinity is

$$f = \frac{1}{s^2}\left[y - \frac{1}{s}(1 - e^{-sy})\right] \tag{8.29}$$

For the axisymmetric stagnation flow on a circular cylinder, use Eqs. (8.7, 8.10) to obtain

$$f = \left[\frac{r^2 - 1}{2} + \frac{rK_1(sr) - K_1(s)}{sK_0(s)}\right] \tag{8.30}$$

where the K's are modified Bessel functions.

8.2.4 D–B Flow Past a Sphere and a Cylinder

For the uniform unsteady D–B flow past a sphere, let $(\varrho, \theta, \varphi)$ be spherical coordinates. Equations (8.7, 8.11) give the stream function

$$\psi = e^{\lambda t}\frac{\sin^2(\theta)}{2}\left[\varrho^2 - \frac{s^2 + 3s + 3}{s^2\varrho} + \frac{3e^{-s(\varrho-1)}}{s^2}\left(s + \frac{1}{\varrho}\right)\right] \tag{8.31}$$

This form is very similar to Eq. (6.65), which is unsteady Stokes flow over a sphere. The difference is that the time exponent in Eq. (8.31) may be accelerating or decelerating, provided the real part of $s = \sqrt{k^2 + \lambda} > 0$, while Eq. (6.65) is always accelerating.

When $\lambda = -k^2$, Eq. (8.31) reduces to

$$\psi = e^{-k^2 t}\frac{\sin^2(\theta)}{2}\left(\varrho^2 + \frac{1}{2\varrho} - \frac{3}{2}\varrho\right) \tag{8.32}$$

which is decelerating, and reduces to the steady Stokes flow solution Eq. (6.27) when $k = 0$.

Oscillatory flow is possible when λ is complex, similar to Sect. 8.2.1.1.

For the cylinder, the solution in cylindrical coordinates is

$$\psi = e^{\lambda t} \sin(\theta) \left[r - \frac{1}{r} - \frac{2K_1(s)}{sK_0(s)r} + \frac{2K_1(sr)}{sK_0(s)} \right] \tag{8.33}$$

The comments are similar to those of the sphere, except the solution does not exist when $s = 0$ or $\lambda = -k^2$.

When $\lambda = 0$ and $s = k \neq 0$, we obtain the uniform steady flow over a circular cylinder in a porous medium. There is no Stokes paradox.

8.3 Solution Using Infinite Series

Since D–B flow is linear, one can superpose any number of closed-form solutions from Sect. 8.2 provided the boundary is the same. For example one can construct a pulse solution from the time dependence and integrate the (Green's function) solution for arbitrary time dependence. In what folllows we shall illustrate with some series solutions through ducts.

8.3.1 Starting D–B Flow in a Rectangular Duct

Consider a rectangular duct filled with a porous medium. A sudden constant pressure gradient is applied. The governing equation is Eq. (8.4). Let

$$w = \overline{w}(x, y) - \tilde{w}(x, y, t) \tag{8.34}$$

where $\overline{w}$ is the steady state solution and $\tilde{w}$ is the transient solution which decays to zero as time increases. For the steady solution, the equation is

$$\nabla^2 \overline{w} - k^2 \overline{w} = -1 \tag{8.35}$$

with the boundary conditions

$$\overline{w}(\pm 1, y) = 0, \quad \overline{w}(x, \pm b) = 0 \tag{8.36}$$

Here the coordinates are at the center of the duct cross section and b is the aspect ratio. Let

$$\overline{w} = \sum_{n=1}^{\infty} \cos(\alpha_n x) f_n(y), \quad \alpha_n = \left(n - \frac{1}{2} \right) \pi, \tag{8.37}$$

Also, expand unity in Fourier series

$$1 = \sum_{n=1}^{\infty} c_n \cos(\alpha_n x), \quad c_n = \frac{2(-1)^{n+1}}{\alpha_n} \tag{8.38}$$

Equation (8.35) becomes

$$f_n'' - \left(k^2 + \alpha_n^2\right) f_n = -c_n \tag{8.39}$$

The solution satisfying Eq. (8.36) is

$$f_n = \frac{c_n}{s^2}\left[1 - \frac{\cosh(sy)}{\cosh(sb)}\right], \quad s = \sqrt{k^2 + \alpha_n^2}. \tag{8.40}$$

For the transient, the equation is

$$\nabla^2 \tilde{w} - k^2 \tilde{w} = \tilde{w}_t \tag{8.41}$$

with the boundary conditions that $\tilde{w}$ is zero on all walls and

$$\tilde{w}(x, y, 0) = \overline{w}(x, y) \tag{8.42}$$

Let the decaying solution be

$$\tilde{w} = \sum_{n=1}^{\infty} \sum_{m=1}^{\infty} A_{mn} \cos(\alpha_n x) \cos(\beta_m y) e^{-\gamma_{mn} t}, \quad \beta_m = \left(m - \frac{1}{2}\right)\frac{\pi}{b} \tag{8.43}$$

Equation (8.41) gives

$$\gamma_{mn} = \alpha_n^2 + \beta_m^2 + k^2 \tag{8.44}$$

Equation (8.42) yields

$$\sum_{m=1}^{\infty} A_{mn} \cos(\beta_m y) = f_n(y) \tag{8.45}$$

Invert for the Fourier coefficients

$$A_{mn} = \frac{2c_n(-1)^{m+1}}{b\beta_m \gamma_{mn}} \tag{8.46}$$

The integrated flow rate depends on the aspect ratio, the porous medium factor and the time. It rises smoothly from zero to the steady state value.

8.3.2 D–B Flow in a Sector Duct

This problem was done by Wang (2010c). Figure 8.2 shows the normalized cross section of a sector duct which can fit between intersecting planes. Let the opening angle be 2β and the solution satisfying the boundary conditions on the straight sides be

Fig. 8.2 The sector duct cross section

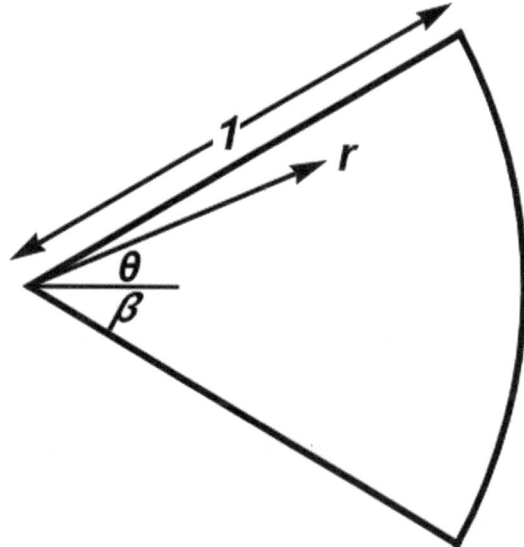

$$w = \sum_{n=1}^{\infty} \cos(\alpha_n \theta) f_n(r), \quad \alpha_n = \left(n - \frac{1}{2}\right)\frac{\pi}{\beta}. \tag{8.47}$$

Expand unity similar to Eq. (8.38). Equation (8.35) gives

$$f_n'' + \frac{1}{r}f_n' - \frac{\alpha_n^2}{r^2}f_n - k^2 f_n = -c_n = \frac{2(-1)^n}{\alpha_n \beta} \tag{8.48}$$

The general bounded solution satisfying no-slip at $r = 1$ is

$$f_n = \sum_{i=1}^{\infty} A_{ni} J_{\alpha_n}(l_{ni}r) \tag{8.49}$$

where l_{ni} is the ith zero of the Bessel function J. Equation (8.48) gives

$$\sum_{n=1}^{\infty} A_{ni}\left(k^2 + l_{ni}^2\right) J_{\alpha_n}(l_{ni}r) = -c_n \tag{8.50}$$

Using the orthogonality of the Bessel function we find

$$A_{ni} = \frac{-2c_n F_{ni}}{\left(k^2 + l_{ni}^2\right) J_{\alpha_n - 1}(l_{ni}) J_{\alpha_n + 1}(l_{ni})} \tag{8.51}$$

Here

$$F_{ni} = \int_0^1 r J_{\alpha_n}(l_{ni} r) dr \tag{8.52}$$

can be expressed in a sum of Bessel functions. The solution for w is then obtained from
Eq. (8.47).

8.4   Point Match Method

The point match method, briefly introduced in Sect. 6.11.2, will be applied to the D–B
equation.

8.4.1   D–B Flow Over a Grooved Surface

Figure 8.3a shows a uniform flow in a porous medium over a grooved surface. The
grooves model roughness or striations. Notice that, due to Stokes paradox, there is no
clear fluid counterpart. Place normalized Cartesian coordinates in a repeated domain,
as shown in Fig. 8.3b. Decompose the domain into Region I $(-1 \leq x \leq 1, y \geq 0)$ and
Region II $(-a \leq x \leq a, -b \leq y \leq 0)$.

Both Regions satisfy

$$\nabla^4 \psi - k^2 \nabla^2 \psi = 0 \tag{8.53}$$

For Region I, the solution which is periodic in x and tends to uniform flow at infinity is

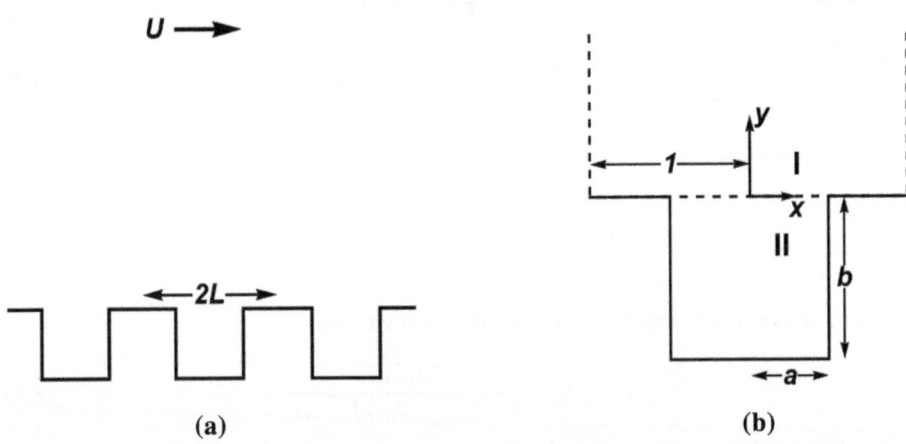

(a)   (b)

Fig. 8.3 **a** D–B flow over a grooved surface **b** domain decomposition of a repeated strip

$$\psi_I(x, y) = y + A_0 + B_0 e^{-ky} + \sum_{n=1}^{N-1} \cos(\alpha_n x)\left(A_n e^{-\alpha_n y} + B_n e^{-\beta_n y}\right) \tag{8.54}$$

where

$$\alpha_n = n\pi, \quad \beta_n = \sqrt{\alpha_n^2 + k^2} \tag{8.55}$$

For Region II, the solution which is zero on the three walls is

$$\psi_{II}(x, y) = \sum_{m=1}^{M} \cos(\gamma_m x)\{C_m \sinh[\gamma_m(y + b)] + D_m \sinh[\delta_m(y + b)] +$$

$$E_m \cosh[\gamma_m(y + b)] - \cosh[\delta_m(y + b)]\} + \sum_{p=1}^{P} \sin(\mu_p y) F_p \left[\frac{\cosh(\mu_p x)}{\cosh(\mu_p a)} - \frac{\cosh(v_p x)}{\cosh(v_p a)}\right] \tag{8.56}$$

where

$$\gamma_m = \left(m - \frac{1}{2}\right)\frac{\pi}{a}, \quad \delta_m = \sqrt{\gamma_m^2 + k^2}, \quad \mu_p = \frac{p\pi}{b}, \quad v_p = \sqrt{\mu_p^2 + k^2} \tag{8.57}$$

The rest of the boundary conditions are no-slip on the walls and matching between the two Regions. The point match method is applied as follows. Let

$$x_i = \frac{(i - 0.5)}{N}, \quad y_j = \frac{(j - 0.5)b}{M} \tag{8.58}$$

The no-slip condition on the side walls is

$$\psi_{IIx}(a, y_j) = 0, \quad j = 1 \text{ to P.} \tag{8.59}$$

The no-slip condition on the bottom wall is.

$$\psi_{IIy}(x_i, -b) = 0, \quad i = 1 \text{ to M} \tag{8.60}$$

Matching conditions on $y = 0$ are

$$\psi_I(x_i, 0) = \begin{cases} \psi_{II}(x_i, 0), & i = 1 \text{ to } M \\ 0, & i = M + 1 \text{ to } N \end{cases} \tag{8.61}$$

$$\psi_{Iy}(x_i, 0) = \begin{cases} \psi_{IIy}(x_i, 0), & i = 1 \text{ to } M \\ 0, & i = M + 1 \text{ to } N \end{cases} \tag{8.62}$$

$$\psi_{Iyy}(x_i, 0) = \psi_{IIyy}(x_i, 0), \quad i = 1 \text{ to } M \tag{8.63}$$

$$\psi_{Iyyy}(x_i, 0) = \psi_{IIyyy}(x_i, 0), \quad i = 1 \text{ to } M \tag{8.64}$$

There are $2N$ equations from Eqs.(8.61, 8.62), $3M$ equations from Eqs.(8.60, 8.63, 8.64) and P equations from Eq. (8.59) for the $2N + 3M + P$ unknowns: $A_n, B_n, C_m, D_m, E_m, F_p$. These linear algebraic equations are easily inverted. Usually the number of points used is proportional to the length it covers. The accuracy can be increased with increased number of points. The resulting streamlines were determined by Wang (2010b).

8.4.2   D–B Flow Over an Array of Cylinders

Figure 8.4a shows the cross section of an infinite square array of circular cylinders embedded in a porous medium (only four shown). The average fluid velocity is U. The repeated domain is bounded by the dashed lines and one fourth of the cylinder (Fig. 8.4b).

The stream function satisfies Eq. (8.53) with the Laplacian in cylindrical coordinates. The general solution that is anti-symmetric at $\theta = 0$ and symmetric at $\theta = \frac{\pi}{2}$ is

$$\psi(r, \theta) = \sum_{n=1}^{2N} \sin(\alpha_n \theta)\left[A_n r^{\alpha_n} + B_n r^{-\alpha_n} + C_n I_{\alpha_n}(kr) + D_n K_{\alpha_n}(kr)\right] \tag{8.65}$$

where $\alpha_n = 2n - 1$ and I, K are modified Bessel functions. Now on the cylinder

$$\psi(b, \theta) = 0 \tag{8.66}$$

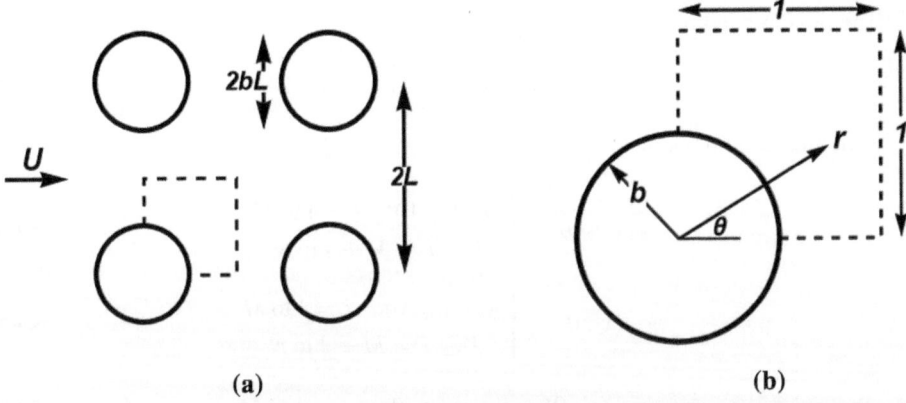

(a)   (b)

Fig. 8.4   **a** Cross section of flow over a square array of cylinders **b** one repeating region

$$\psi_r(b, \theta) = 0 \tag{8.67}$$

On the dashed boundary at $x = 1$, we apply the symmetry condition at evenly spaced points, i.e.,

$$\frac{\partial \psi}{\partial x}(1, y_j) = 0, \quad \frac{\partial^3 \psi}{\partial x^3}(1, y_j) = 0 \tag{8.68}$$

$$y_j = (j - 0.5)/N, \quad j = 1 \text{ to} N \tag{8.69}$$

In cylindrical coordinates, the points are at

$$r_j = \sqrt{1 + y_j^2}, \quad \theta_j = \tan^{-1}(y_j) \tag{8.70}$$

Thus, Eqs.(8.68) are discretized

$$\frac{\partial \psi}{\partial x}\bigg|_{x=1} = \left[\cos(\theta)\frac{\partial}{\partial r} - \frac{\sin(\theta)}{r}\frac{\partial}{\partial \theta}\right]\psi(r_j, \theta_j) \tag{8.71}$$

$$\frac{\partial^3 \psi}{\partial x^3}\bigg|_{x=1} = \left[\cos(\theta)\frac{\partial}{\partial r} - \frac{\sin(\theta)}{r}\frac{\partial}{\partial \theta}\right]^3 \psi(r_j, \theta_j) \tag{8.72}$$

Similarly, on the dashed boundary at $y = 1$, the conditions are

$$\psi(x_i, 1) = 1, \quad \frac{\partial^2 \psi}{\partial y^2}(x_i, 1) = 0 \tag{8.73}$$

$$x_i = (i - 0.5)/N, \quad i = 1 \text{ to} N \tag{8.74}$$

In cylindrical coordinates, the points are at

$$r_i = \sqrt{1 + x_i^2}, \quad \theta_i = \tan^{-1}(1/x_i) \tag{8.75}$$

Equations (8.73) in cylindrical coordinates become

$$\psi(r_i, \theta_i) = 1 \tag{8.76}$$

$$\left[\sin(\theta)\frac{\partial}{\partial r} + \frac{\cos(\theta)}{r}\frac{\partial}{\partial \theta}\right]^2 \psi(r_i, \theta_i) = 0 \tag{8.77}$$

Equations (8.66, 8.67) give $4N$ algebraic equations, Eqs.(8.71, 8.72) $2N$ equations, Eqs.(8.76, 8.77) $2N$ equations. They can be inverted for the $8N$ unknowns A_n, B_n, C_n, D_n without difficulty. See the original work of Wang (2017).

8.5   Ritz Method

As in the point match method, the Ritz method also requires solving of linear algebraic equations, the latter algorithm has now become a built-in library function in any personal computer. The powerful Ritz method has been applied to plates and membranes, but not as often in fluid mechanics.

8.5.1   D–B Flow in a Super-Elliptic Duct

The super-elliptic duct is described by

$$g(x, y) = x^{2n} + \left(\frac{y}{b}\right)^{2n} - 1 = 0 \tag{8.78}$$

where n is a positive integer and b is the aspect ratio. When $n = 1$, the cross section is an ellipse, and when $n \to \infty$ it approaches a rectangle. Between these limits the geometry describes a rectangle with rounded corners (Fig. 8.5).

The D–B equation for parallel flow is

$$\nabla^2 w - k^2 w = -1 \tag{8.79}$$

The boundary condition is that $w = 0$ on the curved wall. Such a curved boundary does not pose any inconvenience for the Ritz method since it is boundary fitted. However, due to the rounded corners, especially for larger n, the problem is unsuitable for point match or numerical integration.

Consider the minimization of the functional J over the cross-sectional area.

$$J = \iint H(w_x, w_y, w)dxdy, \quad H = w_x^2 + w_y^2 + k^2 w^2 - 2w \tag{8.80}$$

Using Euler's equation, we find minimization leads to Eq. (8.79), and thus, the two problems are equivalent. Now express w in a sum

Fig. 8.5 The super-elliptic cross section. From inside: $n = 1, 2, 5$

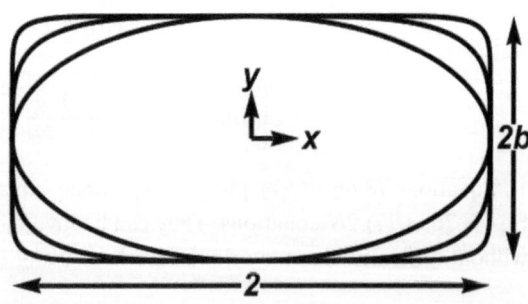

$$w = \sum c_i f_i(x, y). \tag{8.81}$$

where c_i are coefficients to be determined and f_i are Ritz functions which span the space and satisfy the boundary condition. Substitute Eq. (8.81) into Eq. (8.80) and apply the necessary condition for extremum

$$\frac{\partial J}{\partial c_i} = 0 \tag{8.82}$$

The result is an algebraic equation which can be solved for the coefficients c_j

$$\sum_{j=1}^{N} \left(A_{ij} + k^2 B_{ij} \right) c_j = C_i, \quad i = 1 \text{ to } N \tag{8.83}$$

$$A_{ij} = \iint \left(f_{ix} f_{jx} + f_{iy} f_{jy} \right) dx dy, \quad B_{ij} = \iint f_i f_j dx dy, \quad C_i = \iint f_i dx dy \tag{8.84}$$

For the super-elliptic duct, due to symmetry, the Ritz functions can be

$$f_i = g(x, y)\{1, x^2, y^2, x^4, x^2 y^2, y^4, x^6, x^4 y^2, x^2 y^4, y^6, \cdots\} \tag{8.85}$$

Since the sequence is complete, Eq. (8.81) converges to the exact solution as the number of terms N increases (Wang 2010a).

8.5.2   D–B Flow in a Curved Tube

Figure 8.6 shows a curved tube in normalized cylindrical coordinates. Adding a Darcy term in Eq. (1.40) yields the following curved tube equation where the centerline radius and azimuthal pressure gradient have been normalized

$$v_{rr} + \frac{1}{r} v_r + v_{zz} - \frac{v}{r^2} - k^2 v = -\frac{1}{r} \tag{8.86}$$

The cross section of the tube is $g(r, z) = 0$. Let

$$H = r v_r^2 + \frac{v^2}{r} + r v_z^2 - 2v + k^2 r v^2 \tag{8.87}$$

Then, Eq. (8.86) is equivalent to the minimization of

$$J = \iint H \, dz \, dr \tag{8.88}$$

Let $s = r - 1$ and

$$v = \sum c_i f_i(s, z) \tag{8.89}$$

Fig. 8.6 A curved tube cross
section. Dash-dotted line is the
axis of symmetry

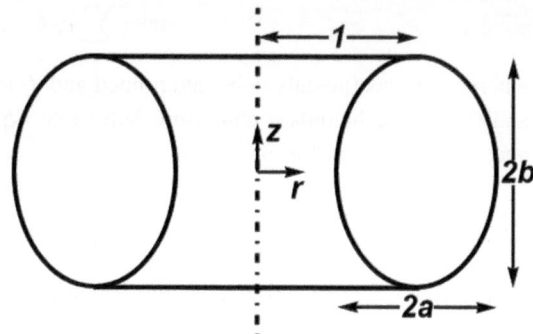

where f_i are the Ritz functions

$$f_i = g\{1, s, s^2, z^2, s^3, sz^2, s^4, s^2z^2, z^4, \cdots\} \tag{8.90}$$

Here, only symmetry with respect to z is assumed. Minimization of J with respect to c_i
gives Eq. (8.83) where

$$A_{ij} = \iint \left[(s+1)(f_{is}f_{js} + f_{iz}f_{jz}) + \frac{f_i f_j}{s+1} \right] dzds$$

$$B_{ij} = \iint (s+1)dzds, \quad C_i = \iint f_i dzds \tag{8.91}$$

The Ritz method can be applied to, for example, an elliptic curved tube given by

$$g = \left(\frac{s}{a}\right)^2 + \left(\frac{z}{b}\right)^2 - 1 \tag{8.92}$$

although our formulation would apply to any cross section symmetric in z. See Wang
(2013).

Exercises

(8.1) Obtain the closed-form solution for unsteady parallel flow in an annulus filled with
a porous medium.

(8.2) Find and discuss the flow due to pressure parallel to a single plate embedded in a
rotating porous medium.

(8.3) Derive the solution for the unsteady stagnation flow on a cylinder embedded in a
porous medium.

(8.4) Formulate and solve the D–B flow over a row of circular cylinders.
(a) the plane of the row is normal to the oncoming flow
(b) the plane of the row is parallel to the oncoming flow (there are two sub-cases)

Is there a Stokes paradox?

Notes

Reviews of flows in porous media can be found in Nield and Bejan (2017), Bear (2018), Wang (2023). Steady closed-form solutions of the D-B equation were reviewed in Wang (2023). Series solution for D-B flow in a sector duct was done by Wang (2010c). Point match method was illustrated by Wang (2010b, 2017). The Ritz method was discussed in Wang (2010a, 2013), Weinstock (1952)

References

J. Bear, *Modelling Phenomena of Flow and Transport in Porous Media* (Springer, Berlin, 2018)

D.A. Nield, A. Bejan, *Convection in Porous Media*, 5th edn (Springer, Berlin, 2017)

C.Y. Wang, Trans. Por. Med. **81**, 207–217 (2010a)

C.Y. Wang, Trans. Por. Med. **84**, 219–227 (2010b)

C.Y. Wang, J. Heat Trans. **132**, 084502 (2010c)

C.Y. Wang, Meccanica **48**, 247–251 (2013)

C.Y. Wang, J. Por. Med. **20**, 749–759 (2017)

C.Y. Wang, J. Por. Med. **26**, 115–123 (2023)

R. Weinstock, *Calculus of Variations* (McGraw-Hill, NY, 1952)

Convection of Heat

9

This chapter considers the close interactions between heat transfer and fluid mechanics. The two most important interactions are free convection and forced convection. In free or natural convection, the temperature differences affect the fluid density, which in turn causes motion through buoyancy. In forced convection, thermal energy is transported by a flowing fluid. This thermal energy may be externally supplied such as through a boundary or internally generated, such as from viscous dissipation. Mixed (free and forced) convection is possible, but not considered here.

The N–S equation, Eq. (1.15), still holds, except in free convection, the body force f includes a buoyancy term

$$f = -g\beta\Delta T \tag{9.1}$$

where g is the gravitational acceleration, β is the coefficient of expansion, and ΔT is the temperature difference. The energy transport equation is

$$T_t + (\boldsymbol{u} \cdot \nabla)T = \kappa\nabla^2 T + \frac{\nu}{c_p}\Phi \tag{9.2}$$

Here, κ is the thermal diffusivity, c_p the specific heat, and Φ is the contribution to heat generation due to viscous stress. From Fourier's law, the heat flux per area is

$$q = -k\frac{\partial T}{\partial \hat{n}} \tag{9.3}$$

where $\hat{n}$ is the normal direction and $k = \rho c_p \kappa$ is the coefficient of conductivity.

Usually, convection problems are approximated empirically or solved numerically. However, analytic solutions can be found under certain restrictions.

Specifically, we shall study external convection where there is always a thermal boundary layer, internal convection where there exists an invariant state independent of axial location, and the onset of fluid motion due to an unstable temperature gradient in porous media.

9.1   External Free Convection

In free convection, the flow is solely due to the temperature difference. Assuming steady flow, the governing equations are

$$(\boldsymbol{u} \cdot \nabla)\boldsymbol{u} = -\frac{1}{\rho}\nabla p + \nu\nabla^2\boldsymbol{u} - \boldsymbol{g}\beta\hat{T} \qquad (9.4)$$

$$(\boldsymbol{u} \cdot \nabla)\hat{T} = \kappa\nabla^2\hat{T} \qquad (9.5)$$

Here, the pressure includes the hydrostatic pressure. $\hat{T}$ is the small temperature deviation from the infinity temperature and affects only the buoyancy term in Eq. (9.4) (Boussinesq approximation).

For external free convection, the velocity and temperature are confined in a boundary layer. Outside the boundary layer almost quiescent conditions exist. In two dimensional Cartesian coordinates, let x be the vertical axis and y normal to it (Fig. 9.1).

Similar to Eq. (4.1), the boundary layer equations are

Fig. 9.1 Free convection boundary layer

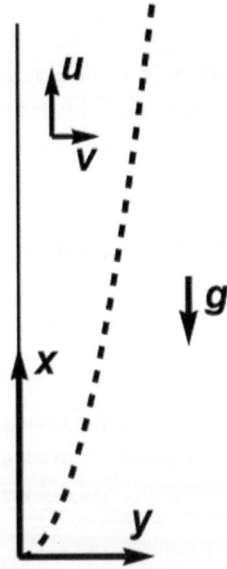

$$\frac{\partial(\psi_y, \psi)}{\partial(x, y)} = \nu\psi_{yyy} + g\beta\hat{T} \tag{9.6}$$

$$\frac{\partial(\hat{T}, \psi)}{\partial(x, y)} = \kappa\hat{T}_{yy} \tag{9.7}$$

9.1.1 Free Convection from a Semi-Infinite Vertical Plate with Constant Temperature

There is no natural length scale, so a similarity solution is possible. Let

$$\psi = x^m f(\eta), \hat{T} = x^\gamma h(\eta), \eta = \frac{y}{x^n} \tag{9.8}$$

Equations (9.6, 9.7) give

$$m + n = 1, \quad \gamma = 1 - 4n \tag{9.9}$$

Since the temperature is constant on the plate at $y = 0$, $x > 0$, the exponents are $\gamma = 0, n = \frac{1}{4}, m = \frac{3}{4}$. Now start afresh with the normalization

$$\psi = 4c\nu x^{3/4} f(\eta), \quad \hat{T} = \hat{T}_0 h(\eta), \quad \eta = c\frac{y}{x^{1/4}} \tag{9.10}$$

Pohlhausen (1921) chose

$$c = \left(\frac{g\beta\hat{T}_0}{4\nu^2}\right)^{1/4} \tag{9.11}$$

although there are a few other choices for the length scale exponent. The similarity equations are

$$f''' + 3ff'' - 2(f')^2 + h = 0 \tag{9.12}$$

$$h'' + 3Pfh' = 0 \tag{9.13}$$

Here, $P = \nu/\kappa$ is the Prandtl number, which measures the relative importance of viscosity to diffusivity. The boundary conditions are

$$f(0) = 0, f'(0) = 0, f'(\infty) = 0, h(0) = 1, h(\infty) = 0. \tag{9.14}$$

These equations can be integrated by a desk computer. The heat loss increases and the induced flow decreases with increased Prandtl number.

9.1.2   Other Free Convection Similarity Solutions

(a) Free convection from a vertical plate heated by constant flux

If the plate is heated by uniformly distributed heating elements, the heat flux can be regarded as constant. Equation (9.3) shows if

$$\frac{\partial \hat{T}}{\partial y} = -\frac{q}{k} \tag{9.15}$$

Equation (9.8) gives $\gamma = n$ or $\gamma = \frac{1}{5}, n = \frac{1}{5}, m = \frac{4}{5}$. One can proceed as in the previous section.

(b) Free convection from a two-dimensional source

This is the plume caused by a heated horizontal wire. The thermal energy is constant for each horizontal cross section.

$$\int_0^\infty \rho c_p u \hat{T} dy = E \tag{9.16}$$

Substituting $u = \psi_y$, Eq. (9.16) gives $m + \gamma = 0$ or $\gamma = \frac{-3}{5}, n = \frac{2}{5}, m = \frac{3}{5}$.

(c) Free convection from a vertical needle

A needle is a slender axisymmetric body (Sect. 4.2.2.2). Using cylindrical coordinates and assuming axisymmetric boundary layer about the z axis, Eq. (1.39), with the buoyancy term and without the pressure gradient, is

$$\frac{1}{r^2}\frac{\partial(\psi, \psi_r)}{\partial(r, z)} + \frac{1}{r^3}\psi_z \psi_r = \nu\left(\frac{1}{r}\psi_{rrr} - \frac{1}{r^2}\psi_{rr} + \frac{1}{r^3}\psi_r\right) + g\beta \hat{T} \tag{9.17}$$

$$\frac{1}{r}\frac{\partial(\psi, \hat{T})}{\partial(r, z)} = \kappa\left(\hat{T}_{rr} + \frac{1}{r}\hat{T}_r\right) \tag{9.18}$$

Applying similarity transform on Eqs. (9.17, 9.18), we find $m = 1$ and $\gamma = 1 - 4n$.

Wang (1989) considered a needle with a heat source at the bottom tip (a stick burning at bottom end). The invariance of thermal energy gives

$$\int_0^\infty \rho c_p w \hat{T} 2\pi r dr = E \tag{9.19}$$

This gives $\gamma + m = 0$. Solving for the three exponents we find $\gamma = -1, m = 1, n = 1/2$. The geometry of the needle is the paraboloid $r = \sqrt{az}$, where a is a small constant. Start afresh with

$$\psi = vzf(\eta), \quad \hat{T} = \frac{4v^2}{g\beta a^2}\frac{h(\eta)}{z}, \quad \eta = \frac{r^2}{az} \tag{9.20}$$

The governing equations are

$$2(\eta f'')' + ff'' + h = 0, \quad 2(\eta h')' + P(fh')' = 0 \tag{9.21}$$

$$f(1) = 0, \quad f'(1) = 0, \quad f'(\infty) = 0, \quad h(\infty) = 0 \tag{9.22}$$

$$\int_1^\infty f'h d\eta = \alpha \equiv g\beta a^2 E/8\pi\rho c_p v^3 \tag{9.23}$$

These equations can be integrated for given Prandtl number P and normalized energy α. Notice conservation of energy Eq. (9.23) implies the needle is adiabatic, $h'(1) = 0$.

9.2   External Forced Convection

In forced convection problems, the momentum equation is first established, and the resulting flow field affects the energy equation through convection or dissipation. We shall present some typical examples for external forced convection in this section.

9.2.1   Shear Flow Over a Plate with Heat Input

Consider a linear shear flow over an infinite plate (Fig. 9.2). The shear flow

$$U = ay \tag{9.24}$$

which could be due to Couette flow or any approximate velocity distribution close to the plate.

Fig. 9.2 Shear flow over a plate. Dashed curve shows the boundary of thermal boundary layer

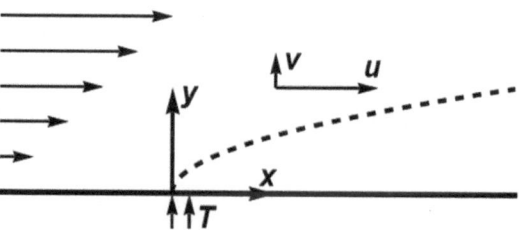

If a power law temperature difference (from that at infinity) is applied on the plate for positive x

$$\hat{T} = \begin{cases} 0, x < 0, y = 0 \\ bx^m, x > 0, y = 0 \end{cases} \tag{9.25}$$

A thermal boundary layer begins at $x = 0$. The forced convection equation is

$$U\hat{T}_x = \kappa\hat{T}_{yy} \tag{9.26}$$

We seek a similarity solution whenever possible. Let

$$\hat{T} = x^m h(\eta), \quad \eta = \frac{y}{x^n} \tag{9.27}$$

Equation (9.26) gives $n = 1/3$ and

$$a\eta\left(mh - \frac{1}{3}\eta h'\right) = \kappa h'' \tag{9.28}$$

One can eliminate the constant parameters by renormalizing

$$\hat{T} = bx^m h(\eta), \quad \eta = \left(\frac{\kappa}{a}\right)^{1/3}\frac{y}{x^n} \tag{9.29}$$

Then

$$\eta\left(mh - \frac{1}{3}\eta h'\right) = h'' \tag{9.30}$$

(a) If the plate is at a constant temperature starting from the origin, $m = 0, b = \hat{T}_0$. The boundary conditions are

$$h(0) = 1, \quad h(\infty) = 0 \tag{9.31}$$

Leveque (1928) found the solution

$$h = \frac{3^{1/3}}{\Gamma\left(\frac{1}{3}\right)}\int_\eta^\infty e^{-\eta^3/9}d\eta \tag{9.32}$$

where Γ is the Gamma function.

(b) If the plate has constant flux input q, Eq. (9.3) gives $m = n = 1/3$. Take

$$b = \frac{q}{k(\kappa/a)^{1/3}}, \quad h(0) = -1 \tag{9.33}$$

Worsoe-Schmidt (1967) found

$$h = \left(\frac{3^{2/3}}{\Gamma\left(\frac{2}{3}\right)}\right)\left(e^{-\eta^3/9} - \frac{\eta}{3}\int_{\eta}^{\infty} \eta e^{-\eta^3/9}d\eta\right) \tag{9.34}$$

(c) If the plate is adiabatic and a heat source is at the origin, E is constant

$$E = \int_0^{\infty} \rho c_p u \hat{T} dy \tag{9.35}$$

Thus $m = -2n = -2/3$. Take

$$b = \frac{E}{\rho c_p a(a/\kappa)^{2/3}}, \quad \int_0^{\infty} \eta h d\eta = 1 \tag{9.36}$$

Wang (1991) found

$$h = \frac{1}{3^{1/3}\Gamma\left(\frac{2}{3}\right)}e^{-\eta^3/9} \tag{9.37}$$

9.2.2 Uniform Flow Over a Heated Semi-infinite Plate

The fluid flow is the Blasius boundary layer, studied in Sect. 4.2.1.1. The convection problem is similar to Fig. 9.2, except the outer velocity is uniform, and the plate exists for $x \geq 0$.

For the velocity boundary layer, let

$$\psi = \sqrt{av}\sqrt{x}f(\eta), \quad \eta = \sqrt{\frac{a}{v}}\frac{y}{\sqrt{x}} \tag{9.38}$$

Equation (4.3), without the pressure gradient, gives the similarity equation

$$-\frac{1}{2}ff'' = ff''' \tag{9.39}$$

If $a = U_{\infty}$, the boundary conditions are $f(0) = 0$, $f'(0) = 0$, $f'(\infty) = 1$. The solution is $f''(0) = 0.332056$ from Table 4.1.

Let the temperature distribution be

$$\hat{T} = cx^{\gamma}h(\eta) \tag{9.40}$$

Equation (9.7) gives

$$P\left(\gamma f'h - \frac{1}{2}fh'\right) = h'', \quad h(\infty) = 0 \tag{9.41}$$

We can set, without loss of generality, $h(0) = 1$. The following are some special cases.

(a) Constant temperature plate

Set $\gamma = 0$ and $c = \hat{T}_0$, which is the temperature difference of the plate. The boundary condition is $h(0) = 1$. Pohlhausen (1921) found the analogy between energy and momentum equations

$$h' = A(f'')^P \tag{9.42}$$

Thus,

$$h = \frac{\int_\eta^\infty (f'')^P \, d\eta}{\int_0^\infty (f'')^P \, d\eta} \tag{9.43}$$

(b) Constant flux plate
Equation (9.3) gives $\gamma = 1/2$ and

$$c = \frac{q}{-k\sqrt{\frac{a}{\nu}}h'(0)} \tag{9.44}$$

For each Prandtl number P, Eq. (9.41) is integrated numerically, and $h'(0)$ found.
(c) Point heat source at the origin.
Equation (9.35) yields $\gamma = -1/2$ and

$$c = \frac{E}{\rho c_p \sqrt{a\nu} \int_0^\infty f'h \, d\eta} \tag{9.45}$$

Equation (9.41) gives

$$-\frac{P}{2}\left(fh' + hf'\right) = h'' \tag{9.46}$$

Wang (2021) integrated this equation twice to obtain

$$h = e^{-\frac{P}{2}\int_0^\eta f \, d\eta} \tag{9.47}$$

9.3   Internal Forced Convection

Forced convection in ducts is the most efficient means for the transport of heat. Heat transfer may occur through the duct walls, aside from internal heat generation due to viscous dissipation. Consider an established steady flow in a long duct. A thermally developing region begins when the heat transfer on the wall first occurs. Further downstream, there exists a fully developed thermal state, where the temperature distribution becomes invariant, or at most linear, with respect to any downstream location.

The thermally developing region, similar to the entrance region for viscous flow, is very tedious analytically aside from some uncertain assumptions. For example, the analytic solution of a thermally developing problem (Graetz problem) involves difficult eigenvalues and eigenfunctions such that solving it numerically becomes more appropriate.

The two most important wall boundary conditions are constant temperature (T problem) and constant flux (H problem). See Shah and London (1978) for a review. In this section we shall consider only hydrodynamically and thermally fully developed analytic solutions.

The Nusselt number is a measure of the importance of convection to conduction, especially relevant in forced convection. It is defined as

$$Nu = \frac{hD_h}{k}, \quad h = \frac{q}{T_s^* - T_m^*} \tag{9.48}$$

Here, h is the convection coefficient, k is the conductivity, $D_h = 4(\text{area})/(\text{perimeter})$ is the hydraulic diameter, q is the mean peripheral heat flux, T_s^* is the mean surface temperature, and T_m^* is the mean bulk temperature defined by

$$T_m^* = \frac{\iint wT^* dA}{\iint w\, dA} \tag{9.49}$$

where w is the axial velocity, and the integration is over the cross sectional area.

9.3.1   Flow in a Circular Tube with Dissipation and Convective Boundary Condition

Normalize lengths by the tube radius L, the velocity by $U = GL^2/\mu$, G being the pressure gradient, the Poiseuille solution in cylindrical coordinates (r, z) is

$$w = \frac{1}{4}\left(1 - r^2\right) \tag{9.50}$$

Normalize the temperature difference (with respect to ambient temperature) by $\mu U^2/k$ and the dissipation function by U^2/L^2. Equation (9.2) becomes

$$\nabla^2 \hat{T} = -\Phi = -(w_r)^2 \tag{9.51}$$

For a wall cooled by convection (or weak radiation), the dimensional boundary condition is

$$h\hat{T} + k\hat{T}_{\hat{n}} = 0 \tag{9.52}$$

where h is the convection coefficient and $\hat{n}$ is the dimensional normal. Equation (9.52) becomes, on $r = 1$,

$$\hat{T} + \alpha\hat{T}_r = 0, \quad \alpha = k/hL \tag{9.53}$$

The solution to Eqs. (9.51, 9.53) is

$$\hat{T} = \frac{1}{64}\left(1 - r^4 + 4\alpha\right) \tag{9.54}$$

Equations (9.48, 9.49) yield the Nusselt number

$$Nu = \frac{48}{5} \tag{9.55}$$

which is independent of location. The Nusselt number, however, will be dependent on α if the cross-section is not circular.

9.3.2 H1 Constant Flux Convection in a Rounded Triangular Duct

Rounded triangular ducts are used in heat exchanger matrices. The cross section is described by

$$g(x, y) = (1 + m)\left(1 - x^2 - y^2\right) + \frac{m}{3}\left(1 - x^3 + 3xy^2\right) = 0 \tag{9.56}$$

When $m = 0$, it is a circle and when $m = \infty$, it becomes an equilateral triangle. For values in between, the cross section is a rounded triangle as shown in Fig. 9.3.

This shape leads to a rare, closed-form solution for viscous flow in a duct. Normalize lengths by the inscribing radius L and velocity by $U = GL^2/\mu$, as in the previous section. The N–S equation reduces to

$$\frac{\partial^2 w}{\partial x^2} + \frac{\partial^2 w}{\partial y^2} = -1 \tag{9.57}$$

The solution is exact

$$w = \frac{1}{4(1 + m)} g(x, y) \tag{9.58}$$

Fig. 9.3 The rounded
triangular cross section. From
inside: $m = 0, 5, 100$

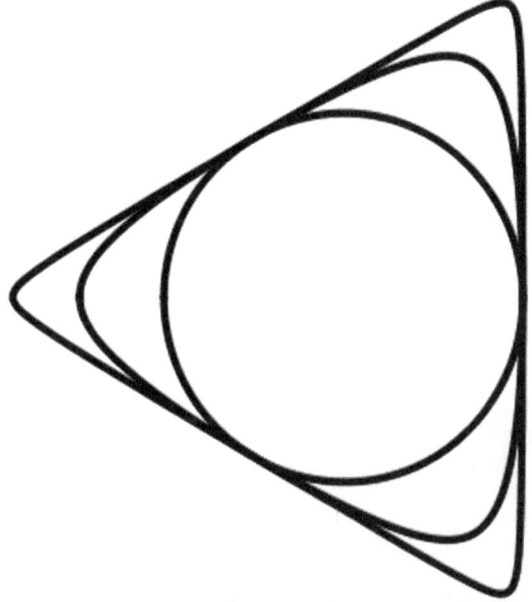

Now consider constant flux heat input on the wall. Assume the resulting wall temperature is constant peripherally but linearly increasing axially (H1 problem). Let $T^*(x, y, z)$ be the unknown temperature and $T_m^*(z)$ be the mean temperature. An energy balance on an elemental segment of the duct yields

$$\iint w \frac{\partial T^*}{\partial z} dA = Q \frac{dT_m^*}{dz}, \quad Q = \iint w \, dA \tag{9.59}$$

The energy equation without dissipation and axial diffusion is

$$\frac{GL}{\mu} w \frac{\partial T^*}{\partial z} = \frac{\kappa}{L^2} \left(\frac{\partial^2 T^*}{\partial x^2} + \frac{\partial^2 T^*}{\partial y^2} \right) \tag{9.60}$$

Define τ as the normalized temperature deviation from the peripheral temperature

$$T^* = T_s^*(z) + \frac{qLP}{kQ} \tau(x, y) \tag{9.61}$$

where P is the peripheral length of the cross section. Since all temperature must linearly increase axially,

$$\frac{\partial T^*}{\partial z} = \frac{dT_m^*}{dz} = \frac{dT_s^*}{dz} = \frac{\kappa \mu q P}{GL^2 kQ} \equiv C \tag{9.62}$$

Equation (9.60) gives

$$\frac{\partial^2 \tau}{\partial x^2} + \frac{\partial^2 \tau}{\partial y^2} = w \tag{9.63}$$

with the boundary condition $\tau = 0$ on the wall. The solution is

$$\tau = \frac{\left[(1+m)(x^2 + y^2) - 3 - 4m\right]}{64(1+m)^2} g(x, y) \tag{9.64}$$

Thus, $T_s^* = Cz + T_a^*$, where T_a^* is the starting (ambient) temperature at $z = 0$. Equation (9.61) then gives the complete temperature distribution.

Equations (9.59, 9.61) yield

$$T_m^* = T_s^* + \frac{qPL}{kQ^2} S \tag{9.65}$$

where

$$S = \iint w\tau dA \tag{9.66}$$

is an energy integral. From Eq. (9.48), the Nusselt number is

$$Nu = \frac{4AQ^2}{P^2 |S|} \tag{9.67}$$

For the circular duct $m = 0$, the Nusselt number is 48/11. For the equilateral triangle, the Nusselt number is 28/9. The results for the rounded triangular duct are found exactly by Wang (2010).

9.3.3 Ritz Method for H2 Forced Convection

In constant heat flux problems, the wall temperature for the H1 problem is constant peripherally but increases axially. In the H2 problem the peripheral wall temperature varies due to locally applied flux. Thus, the H1 problem models highly conductive walls such as metal, and the H2 problem models poorly conductive walls such as ceramic. For the circular duct or for parallel plates, the H1 or H2 solutions are identical.

The temperature equations for the H2 problem are similar to those of the previous section, except the boundary condition $\tau = 0$ is replaced by the local flux

$$\frac{\partial \tau}{\partial n} = \frac{Q}{P} \equiv c \tag{9.68}$$

An additional condition for the Neumann condition is

$$\oint \tau dP = 0 \tag{9.69}$$

For geometries without exact solutions, a Ritz method can be used for the viscous flow in a duct, as presented in Sect. 8.5, and a similar method can be constructed for the H1 forced convection. The Ritz method for the H2 convection is slightly more involved and described as follows.

Consider the functional

$$K = \iint \left(\tau_x^2 + \tau_y^2 + 2w\tau \right) dA - 2c \oint \tau dP \tag{9.70}$$

Using variational calculus, the minimization of K results in

$$\iint (\tau_{xx} + \tau_{yy} - w)\delta \tau dA + \oint \left(\frac{\partial \tau}{\partial n} - c \right) \delta \tau dP = 0 \tag{9.71}$$

Now if τ is arbitrary inside the region and also on the boundary, the parentheses in Eq. (9.71) must be zero, which give Eqs. (9.63, 9.68). Express

$$\tau = b_0 + \sum_{i=1}^{N} b_i g_i \tag{9.72}$$

where g_i are the Ritz functions without any restrictions except for symmetry properties of the cross section and the absence of a constant term. Minimization of K with respect to the coefficients gives

$$\frac{\partial K}{\partial b_i} = 0 \tag{9.73}$$

or

$$\sum_{j=1}^{N} b_j D_{ij} = cE_i - F_i, \quad i = 1 \text{ to } N \tag{9.74}$$

where

$$D_{ij} = \iint \left(g_{ix} g_{jx} + g_{iy} g_{jy} \right) dA, \quad E_i = \oint g_i dP, \quad F_i = \iint w g_i dA \tag{9.75}$$

Equation (9.75) is inverted for the N coefficients b_i. The coefficient b_0 is found from Eq. (9.69)

$$b_0 = \frac{-1}{P} \sum_{i=1}^{N} b_i E_i \tag{9.76}$$

The details can be found in Wang (2014).

9.4   Thermo-Convective Stability in Porous Media

The onset of instability in a bottom-heated fluid-saturated porous medium is important for heat transfer in porous rock and insulation. When a critical Rayleigh number is reached, convection begins, and the heat transfer increases dramatically.

Figure 9.4 shows the cross section of a layer of porous medium. The bottom temperature T_1^* is larger than the top temperature T_0^*. For pure conduction, the temperature is linear

$$T^* = T_1^* - \Delta T^* \frac{z^*}{L} \tag{9.77}$$

where $\Delta T^* = T_1^* - T_0^*$.

The Darcy-Boussinesq equation is

$$\boldsymbol{u}^* = -\frac{K}{\mu}\left[\nabla p^* - \rho_0 g \beta \left(T^* - T_0^*\right)\right] \tag{9.78}$$

where K is the permeability, ρ_0 is the density at T_0^*, g is the gravitational acceleration in the negative z^* direction, and β is the coefficient of expansion. Notice that even without the Brinkman term, viscosity μ still manifests. The energy transport equation is

$$\left(\boldsymbol{u}^* \cdot \nabla\right)T^* = \kappa \nabla^2 T^* \tag{9.79}$$

Here, κ is the effective thermal diffusivity. The continuity equation is

$$\nabla \cdot \boldsymbol{u}^* = 0 \tag{9.80}$$

Fig. 9.4 A porous layer

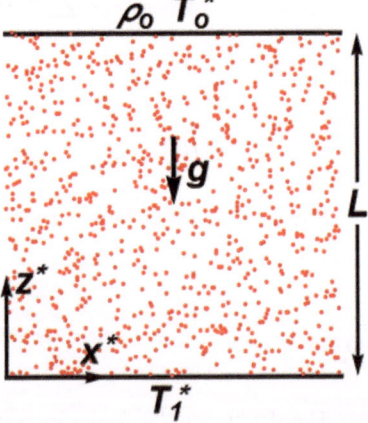

Normalize all lengths by L, the velocities by $\frac{\kappa}{L}$, the pressure by $\frac{\kappa\mu}{K}$, the temperature by ΔT^*. Let w be the normalized velocity in the z direction. Equation (9.78) gives

$$w = -p_z + R(T - T_0) \tag{9.81}$$

where R is the Rayleigh number

$$R = \frac{LK\rho_0 g\beta\Delta T^*}{\mu\kappa} \tag{9.82}$$

Perturb the normalized Eq. (9.77)

$$T = T_1 - z + \tau, \quad \tau \ll 1 \tag{9.83}$$

Equations (9.78, 9.80) yield

$$\nabla^2 p = RT_z \tag{9.84}$$

Now take the Laplacian on Eq. (9.81) to eliminate p, and use Eq. (9.83) to obtain

$$\nabla^2 w = R(\nabla^2\tau - \tau_{zz}) = R\nabla_1^2\tau \tag{9.85}$$

where ∇_1^2 is the Laplace operator in the horizontal directions only. The leading order of the normalized Eqs. (9.79) and (9.83) gives

$$-w = \nabla^2\tau \tag{9.86}$$

Eliminating w from Eqs. (9.85, 9.86) yields

$$\nabla^4\tau + R\nabla_1^2\tau = 0 \tag{9.87}$$

The boundary conditions are that on top and bottom there is no temperature variation and no penetration

$$\tau = 0, \quad w = 0 \tag{9.88}$$

and that the lateral boundaries (if any) are insulated and without shear stress

$$\frac{\partial\tau}{\partial\hat{n}} = 0, \quad \frac{\partial w}{\partial\hat{n}} = 0 \tag{9.89}$$

The latter condition is due to the absence of the Brinkman term. Equations (9.87–9.89) represent an eigenvalue problem. The smallest eigenvalue, the critical Rayleigh number or normalized temperature difference, gives the instability condition.

Equation (9.87) and the Laplacian in z show the eigenfunction must be in the form

$$\tau = \sin(\pi z) f(x, y) \tag{9.90}$$

Equation (9.87) becomes two-dimensional

$$\left[\left(\nabla_1^2 - \pi^2\right)^2 + R\nabla_1^2\right]f = 0 \tag{9.91}$$

The boundary conditions are that on the lateral walls

$$\frac{\partial f}{\partial \hat{n}} = 0, \quad \frac{\partial}{\partial \hat{n}}\nabla_1^2 f = 0 \tag{9.92}$$

Consider the Helmholtz equation

$$\nabla_1^2 \varphi + \lambda \varphi = 0 \tag{9.93}$$

with Neumann boundary conditions

$$\frac{\partial \varphi}{\partial \hat{n}} = 0 \tag{9.94}$$

Notice that Eq. (9.94) also implies the normal derivative of $\nabla_1^2 \varphi$ is zero. It can be shown the eigenfunctions φ are complete, so that any function with the same boundary conditions can be expressed in these eigenfunctions

$$f = \sum_{i=1}^{\infty} b_i \varphi_i \tag{9.95}$$

Here, φ_i is the ith eigenfunction corresponding to the ith eigenvalue λ_i. Substitution of Eq. (9.95) into Eq. (9.91) yields

$$\sum_{i=1}^{\infty} b_i \left[\left(\lambda_i + \pi^2\right)^2 - R\lambda_i\right] = 0 \tag{9.96}$$

Since f is nontrivial, b_i cannot be all zero, say at least one $b_n \neq 0$. Then the brackets must be zero, or

$$R = \frac{\left(\lambda_n + \pi^2\right)^2}{\lambda_n}. \tag{9.97}$$

The critical Rayleigh number is the minimum of Eq. (9.97) for all possible n or mode shape.

9.4.1   The Two-Dimensional Box

As an illustration consider the two-dimensional box confined by $0 \le z \le 1, 0 \le x \le a$. The Helmholtz eigenfunctions and eigenvalues are

$$\varphi_n = \cos\left(\sqrt{\lambda_n}x\right), \quad \lambda_n = \frac{n^2 \pi^2}{a^2} \tag{9.98}$$

Equation (9.97) gives

$$R = \frac{\pi^2 \left(\frac{n^2}{a^2} + 1\right)^2}{n^2/a^2} \tag{9.99}$$

Equation (9.99) is plotted in Fig. 9.5. The lowest mode is one cell, $n = 1$, for $0 < a < 1.414$. Then two cells for $1.414 < a < 2.450$, three cells for $2.450 < a < 3.464$, etc. The critical Rayleigh number is infinite as $a \to 0$, and tends to $4\pi^2 = 39.478$ for infinite width (Horton and Rogers 1945). Local minima occur at integer a, and local maxima at the intersection of adjacent modes.

The stability analyses for three-dimensional geometries and various boundary conditions are also possible. For literature search the key words are "onset" and "thermo-convective stability".

Exercises

(9.1) Obtain the similarity equations for an axisymmetric plume from a point source.
(9.2) Obtain the similarity equations for a needle with constant temperature.
(9.3) Find the solution for the fully-developed flow and temperature in a rectangular duct with internal dissipation.
(9.4) Find the Nusselt numbers for H1, H2 heat transfer in an equilateral triangular duct.

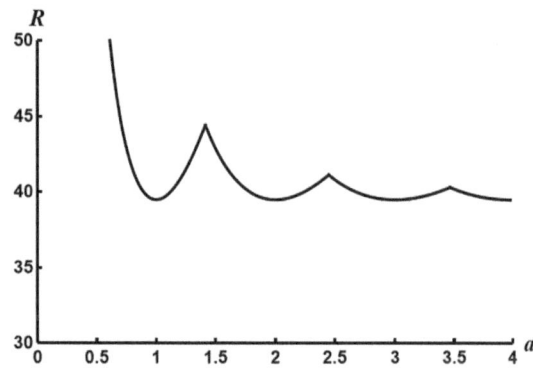

Fig. 9.5 Critical Rayleigh number as a function of width a

(9.5) Solve for the critical Rayleigh number in a layer of saturated porous medium heated from below by constant flux. What if the layer is laterally constrained?

Notes

Good references on convection heat transfer are Bejan (2013), Schlichting and Gersten (2000), Yih (1977). References for external convection are Pohlhausen (1921), Leveque (1928), Worsoe-Schmidt (1967), Wang (1989, 1991, 2021). Best source for internal forced convection is Shah and London (1978). The exact solution for rounded triangular duct was found by Wang (2010). H2 forced convection by the Ritz method was discussed in Wang (2014). Thermo-convective instability was discussed in Horton and Rogers (1945), Nield and Bejan (2017)

References

A. Bejan, *Convection Heat Transfer*, 4th edn (Wiley, NY, 2013)
C.W. Horton, F.T. Rogers, J. Appl. Phys. **16**, 367–370 (1945)
M.A. Leveque, Ann. Des. Mines. **13**, 201–299 (1928)
D.A. Nield, A. Bejan, *Convection in Porous Media*, 5th edn (Springer, NY, 2017)
E. Pohlhausen, ZAMM **1**, 115–121 (1921)
H. Schlichting, K. Gersten, *Boundary Layer Theory*, 8th edn (Springer, Berlin, 2000)
R.K. Shah, A.L. London, *Laminar Flow Forced Convection in Ducts* (Acad. Press, NY, 1978)
C.Y. Wang, Mech. Res. Comm. **16**, 95–101 (1989)
C.Y. Wang, J. Heat Trans. **113**, 496–498 (1991)
C.Y. Wang, J. Heat Trans. **143**, 094502 (2021)
C.Y. Wang, ZAMM **90**, 522–527 (2010)
C.Y. Wang, J. Thermophys. Heat Trans. **28**, 811–815 (2014)
P.M. Worsoe-Schmidt, Int. J. Heat Mass Trans. **10**, 541–551 (1967)
C.S. Yih, *Fluid Mechanics* (West River Press, MI, 1977)

Appendix A

Modified Stretching Method-Similarity

A similarity transform of a problem decreases the number of independent variables, and often turns a partial differential equation into an ordinary differential equation. The latter is much easier to analyze or solve, analytically or numerically.

The general similarity method utilizes local Lie groups which leaves the differential equations invariant under stretching, translation, rotation and shear transforms. However, for the N–S equation the group theoretic method results in complicated solutions of space and time, which has little physical significance. The more useful similarity solutions of the N–S equation can be obtained by the simpler stretching transform. In the following, we shall present an even simpler version, the modified stretching method. Almost all similarity solutions of the N–S equation can be obtained by this method, for problems where a physical length scale is absent.

The method uses stretching transform of the form

$$\psi = x^m f(\eta), \quad \eta = y/x^n \tag{A.1}$$

so that the N–S equation reduces to an ordinary differential equation.

Example A.1 Consider an example of a plate suddenly moving in its own plane with velocity U (Sect. 2.1.4). The N–S equation is

$$\frac{\partial u}{\partial t} = \nu \frac{\partial^2 u}{\partial y^2} \tag{A.2}$$

There is no physical length scale. Let

$$u = t^m f(\eta), \quad \eta = y/t^n \tag{A.3}$$

© The Editor(s) (if applicable) and The Author(s), under exclusive license
to Springer Nature Switzerland AG 2024
C. Y. Wang, *Essential Analytic Laminar Flow*, Synthesis Lectures on Engineering,
Science, and Technology, https://doi.org/10.1007/978-3-031-36449-5

Equation (A.2) becomes

$$t^{m-1}\left(mf - n\eta f'\right) = vt^{m-2n} f''$$ (A.4)

In order an ordinary differential equation results, we need

$$n = \frac{1}{2}$$ (A.5)

Then the similarity equation is

$$mf - \frac{\eta f'}{2} = v f''$$ (A.6)

The initial and boundary conditions show

$$m = 0, \quad f(0) = U, \quad f(\infty) = 0$$ (A.7)

Example A.2 Consider two-dimensional steady flow. The N–S equation (without body force) in Cartesian coordinates is Eq. (1.21)

$$\frac{\partial \nabla^2 \psi}{\partial x} \frac{\partial \psi}{\partial y} - \frac{\partial \nabla^2 \psi}{\partial y} \frac{\partial \psi}{\partial x} = v \nabla^4 \psi$$ (A.8)

Using Eq. (A.1) the Laplacian is

$$\nabla^2 \psi = x^{m-2}\left[m(m-1)f + (1-2m)n\eta f' + n^2\eta\left(\eta f'\right)'\right] + x^{m-2n} f''$$ (A.9)

In order to have the Laplacian homogeneous in x, there are two non-trivial choices.

(1) $n = 1$, m arbitrary. Substitute

$$\psi = x^m f(\eta), \quad \eta = y/x$$ (A.10)

into Eq. (A.8) and find $m = 0$. This means

$$\psi = f\left(\frac{y}{x}\right)$$ (A.11)

which leads to a similarity equation. The streamlines are rays from the origin. The solution is the Jeffrey-Hamel exact solution (Sect. 2.2.4).

(2) $n = 0$, $m = 1$. Thus

$$\psi = xf(y)$$ (A.12)

Depending on the boundary conditions the N–S equation yields Hiemenz stagnation flow or Crane's stretching flow (Sects. 2.2.1, 2.2.2).

Example A.3 Consider the two-dimensional unsteady N–S equation

$$\frac{\partial \nabla^2 \psi}{\partial t} + \frac{\partial \nabla^2 \psi}{\partial x}\frac{\partial \psi}{\partial y} - \frac{\partial \nabla^2 \psi}{\partial y}\frac{\partial \psi}{\partial x} = \nu \nabla^4 \psi \tag{A.13}$$

Try an extension to Eq. (A.12) as follows

$$\psi = \frac{x}{t^p} f\left(\frac{y}{t^q}\right) \tag{A.14}$$

We find Eq. (A.13) becomes an ordinary differential equation when

$$p = q = \frac{1}{2} \tag{A.15}$$

A further extension replaces t by $1 - \alpha t$, which includes both accelerating and decelerating cases.

$$\psi = \frac{x}{\sqrt{1 - \alpha t}} f\left(\frac{y}{\sqrt{1 - \alpha t}}\right) \tag{A.16}$$

This is the basis for Yang's unsteady stagnation flow (Sect. 2.2.3).

Notes

Aside from the N–S equation, the stretching transform Eq. (A1) can also be applied to the boundary layer equation in Chap. 4. A good source for local Lie group transform and the method of stretching is Logan (1987).

References

J.D. Logan, *Applied Mathematics* (Wiley, NY, 1987)

Appendix B

Perturbation Methods

Perturbation methods are useful when the problem deviates a little bit from a known solution. This chapter is a short introduction to the method.

B.1 Asymptotic Expansion

Consider $f(\varepsilon)$, a function of the small parameter $0 < \varepsilon \ll 1$. We say $f(\varepsilon)$ is the same asymptotic order as $\delta(\varepsilon)$ as $\varepsilon \to 0$, or

$$f(\varepsilon) = O(\delta(\varepsilon)) \tag{B.1}$$

if

$$\lim_{\varepsilon \to 0} \frac{f(\varepsilon)}{\delta(\varepsilon)} = c \tag{B.2}$$

where c is a constant, neither 0 or ∞.

When c is zero in Eq. (B.2), we say $f(\varepsilon)$ is smaller asymptotic order than $\delta(\varepsilon)$

$$f(\varepsilon) = o(\delta(\varepsilon)) \tag{B.3}$$

The function $\delta(\varepsilon)$ is a gauge function, usually powers of ε.

If Eqs. (B.1 and B.2) are true, then as $\varepsilon \to 0$

$$f(\varepsilon) = c\delta(\varepsilon) + o(\delta(\varepsilon)) \tag{B.4}$$

For example

$$\sin(3\varepsilon) = O(\varepsilon), \quad \sin(3\varepsilon) = 3\varepsilon + o(\varepsilon) \tag{B.5}$$

© The Editor(s) (if applicable) and The Author(s), under exclusive license
to Springer Nature Switzerland AG 2024
C. Y. Wang, *Essential Analytic Laminar Flow*, Synthesis Lectures on Engineering,
Science, and Technology, https://doi.org/10.1007/978-3-031-36449-5

For higher order approximations, construct the difference

$$f(\varepsilon) - c_1\delta_1(\varepsilon) \sim c_2\delta_2(\varepsilon) \tag{B.6}$$

where $\delta_2(\varepsilon) = o(\delta_1(\varepsilon))$. Extension to N terms yields

$$f(\varepsilon) = \sum_{n=1}^{N} c_n\delta_n(\varepsilon) + o(\delta_N(\varepsilon)) \tag{B.7}$$

Given the sequence of gauge functions $\delta_m(\varepsilon)$, the coefficients c_n can be *uniquely* determined by the constructive formula

$$c_n = \lim_{\varepsilon \to 0} \frac{f(\varepsilon) - \sum_{m=1}^{n-1} c_m\delta_m(\varepsilon)}{\delta_n(\varepsilon)} \tag{B.8}$$

Notice

- If two asymptotic expansions are equal, all corresponding terms are equal. This can be shown by taking successive limits. Thus, one can equate equal powers of ε in an asymptotic equation.
- Asymptotic expansions are unique. One can obtain the asymptotic expansion by any easier method, such as Taylor series.
- The error of an asymptotic expansion is of O(next term) or o(last term).

Let a function of a small parameter ε be continuous for all $\varepsilon \geq 0$. Then the Taylor series is

$$f(\varepsilon) = f(0) + \varepsilon f'(0) + \frac{\varepsilon^2}{2!} f''(0) + \frac{\varepsilon^3}{3!} f'''(0) + \cdots \tag{B.9}$$

Some Taylor series useful in asymptotic expansions are

- $\sin(x) = x - \dfrac{x^3}{3!} + \dfrac{x^5}{5!} + O(x^7)$
- $\cos(x) = 1 - \dfrac{x^2}{2!} + \dfrac{x^4}{4!} + O(x^6)$
- $e^x = 1 + x + \dfrac{x^2}{2!} + \dfrac{x^3}{3!} + O(x^4)$
- $(1+x)^n = 1 + nx + \dfrac{n(n-1)}{2!}x^2 + O(x^3)$ \hfill (B.10)

B.2 Regular Perturbation

The method is to express the unknown in an asymptotic series of the small parameter and solve successively for each order.

Consider the nonlinear differential equation

$$u''(x) + \varepsilon u u' = u, \quad u(0) = 1, \quad u(\infty) = 0 \tag{B.11}$$

Expand the unknown as

$$u = u_0(x) + \varepsilon u_1(x) + \cdots \tag{B.12}$$

Substituting into Eq. (B.11) and equating equal powers of ε yield

$$u_0''(x) = u_0, \quad u_0(0) = 1, \quad u_0(\infty) = 0 \tag{B.13}$$

$$u_1''(x) + u_0 u_0' = u_1, \quad u_1(0) = 0, \quad u_1(\infty) = 0 \tag{B.14}$$

The solutions for each order are

$$u_0 = e^{-x}, \quad u_1 = \frac{1}{3}\left(e^{-2x} - e^{-x}\right) \tag{B.15}$$

Thus, the asymptotic solution is

$$u = e^{-x} + \frac{\varepsilon}{3}\left(e^{-2x} - e^{-x}\right) + O\left(\varepsilon^2\right) \tag{B.16}$$

B.3 Singular Perturbation

Singular means the small parameter multiplies the highest derivative, such that the zeroth order equation is degenerate (lower order derivatives). The zeroth order solution is the outer solution which should be matched by an inner (boundary layer) solution to satisfy the condition on the boundary.

B.3.1 Boundary Layer Theory

Let us illustrate by an example. Consider the singular problem (not quite the same as Eq. (B.11))

$$\varepsilon u''(x) + u u' = u, \quad u(0) = 1, \quad u(\infty) = 0 \tag{B.17}$$

Using the expansion Eq. (B.12) the outer equation is

$$u_o u_o' = u_o \tag{B.18}$$

The solution is

$$u_o = 0, \quad u_o = x + c \tag{B.19}$$

Only the first solution satisfies the boundary condition at infinity. In order to bring up the importance of the second derivative, let

$$x = \varepsilon \eta \tag{B.20}$$

where η is a stretched variable of order one. The leading terms of Eq. (B.17) yield the boundary layer (inner) equation

$$u_{i\eta\eta} + u_i u_{i\eta} = 0 \tag{B.21}$$

Integrate once to obtain

$$u_{i\eta} + \frac{u_i^2}{2} = c_1 \tag{B.22}$$

Here the integration constant c_1 is set to zero, since u_i must match (mesh) with the zero outer solution. Separating variables and integrating Eq. (B.22) together with the boundary condition $u(0) = 1$ give

$$u_i = \frac{2}{\eta + 2} \tag{B.23}$$

The approximate (boundary layer) solution is

$$u = u_i + u_o - c.p. = \frac{2}{\eta + 2} + 0 + 0 = \frac{2}{\frac{x}{\varepsilon} + 2} + O(\varepsilon) \tag{B.24}$$

Here $c.p.$ denotes the common part to u_i and u_o, which is zero in this example. Notice the algebraic decay and the boundary layer of order ε, where most of the changes take place.

B.3.2 Matched Asymptotic Expansions

This method is also called the method of "inner and outer expansions". It gives the higher order corrections for boundary layers. The five steps are.

(1) Use the zeroth order boundary layer theory to find where and how thick the boundary layer is.
(2) Outer expansion is obtained by expanding the equation directly and satisfying its own boundary condition.
(3) Inner expansion is obtained by expanding the equation in the stretched variable and satisfying its own boundary condition.
(4) The two expansions are matched in an intermediate region.
(5) A composite solution (uniformly valid solution) is constructed by adding the inner and outer solutions and subtracting the common part.

For example, consider the singular problem

$$\varepsilon u'' + 2u' - u^2 = 0, \quad u(0) = 0, \quad u(1) = 1 \tag{B.25}$$

This nonlinear problem has no analytic solution.

(1) First use the simpler boundary layer theory. The outer solution is

$$2u_o' - u_o^2 = 0 \tag{B.26}$$

If the boundary layer is at $x = 0$, then $u_o(1) = 1$ and the solution is

$$u_o = \frac{2}{3 - x} \tag{B.27}$$

In order to bring up the importance of the second derivative, let $x = \varepsilon \eta$. The boundary layer equation is

$$u_{i\eta\eta} + 2u_{i\eta} = 0 \tag{B.28}$$

The solution that is zero at $x = 0$ and meshes into the outer solution is

$$u_i = \frac{2}{3}\left(1 - e^{-2\eta}\right) \tag{B.29}$$

If we assumed the boundary layer is at $x = 1$, a similar process shows the boundary layer solution cannot match any outer solution as $\eta \to -\infty$ (towards the outer region).

(2) For the higher order corrections, let the outer expansion be

$$u_o = f_0(x) + \varepsilon f_1(x) + \cdots \tag{B.30}$$

The solution f_0 is the same as Eq. (B.27). The equation for f_1 is

$$f_0'' + 2f_1' - 2f_0 f_1 = 0, \quad f_1(1) = 0 \tag{B.31}$$

with the solution

$$f_1 = \frac{2}{(3-x)^2} \ln \left[\frac{(3-x)}{2} \right] \tag{B.32}$$

(3) The inner expansion is

$$u_i = g_0(\eta) + \varepsilon g_1(\eta) + \cdots \tag{B.33}$$

The solution g_0 is exactly Eq. (B.29). The next order equation is

$$g_{1\eta\eta} + 2g_{1\eta} - g_0^2 = 0, \quad g_1(0) = 0 \tag{B.34}$$

The solution is

$$g_1 = \frac{1}{9} \left(2\eta + 4\eta e^{-2\eta} + \frac{e^{-4\eta} - 1}{2} \right) + c_1 \left(1 - e^{-2\eta} \right) \tag{B.35}$$

(4) Matching inner and outer solutions

$$u_i|_{\eta \to \infty} = u_o|_{x \to \varepsilon\eta} \tag{B.36}$$

or

$$\frac{2}{3} + \varepsilon \left[\frac{1}{9} \left(2\eta - \frac{1}{2} \right) + c_1 \right] + \cdots = \frac{2}{3 - \varepsilon\eta} + \varepsilon \frac{2}{(3 - \varepsilon\eta)^2} \ln \left[\frac{(3 - \varepsilon\eta)}{2} \right] + \cdots \tag{B.37}$$

Expanding the right-hand side yields

$$c_1 = \frac{2}{9} \ln \left(\frac{3}{2} \right) \tag{B.38}$$

Fig. B.1 The solution of Eq. (B.25) with $\varepsilon = 0.2$. Dotted curve: boundary layer solution with $O(\varepsilon)$ error. Dashed curve: matched asymptotic solution with $O\left(\varepsilon^2\right)$ error. Red continuous curve: Numerical solution of Eq. (B.25).

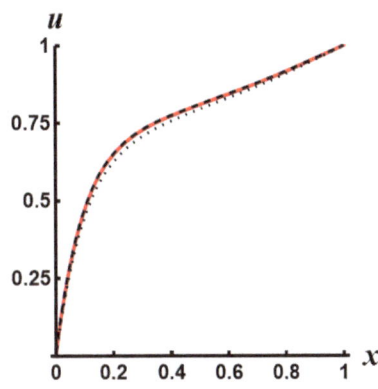

(5) The composite solution is

$$u = \frac{2}{3-x} - \frac{2}{3}e^{-2\eta} + \varepsilon \frac{2}{(3-x)^2}\ln\left[\frac{(3-x)}{2}\right] + \varepsilon\left[\frac{4\eta}{9}e^{-2\eta} + \frac{1}{18}e^{-4\eta} - \frac{2}{9}\ln\left(\frac{3}{2}\right)e^{-2\eta}\right] + O\left(\varepsilon^2\right)$$

(B.39)

Figure B.1 shows a comparison. The boundary layer solution compares fairly well with the numerical solution of Eq. (B.25). But the higher order solution Eq. (B.29) is indistinguishable to the numerical solution.

Notes

A very good book on perturbation methods is Nayfeh (1981). Application of perturbation methods to fluid mechanics is presented in Van Dyke (1964). A more concise book on perturbation methods is Wang (2023)

References

A.H. Nayfeh, *Introduction to Perturbation Techniques* (Wiley, NY, 1981)

M. Van Dyke, *Perturbation Methods in Fluid Mechanics* (Academic, NY, 1964)

C.Y. Wang, *Essential Perturbation Methods* (Springer, Switzerland, 2023)

Appendix C

Potential Flow

Potential flow is inviscid irrotational flow, thus not a primary topic of this book. However, potential flow serves as the outer flow for boundary layer problems, both steady and unsteady. It also governs porous media Darcy flow. We shall limit the discussion of potential flow to the topics which are more applicable to boundary layer problems. For example, in this brief presentation, we shall exclude potential flows over sharp corners where separation usually occurs.

A flow is irrotational if vorticity, defined as the cross product of velocity, remains zero everywhere. This is true for inviscid flow where vorticity is conserved.

$$\zeta = \nabla \times \boldsymbol{u} = \boldsymbol{0} \tag{C.1}$$

Equation (C.1) shows there exists a velocity potential ϕ such that

$$\boldsymbol{u} = \nabla \phi \tag{C.2}$$

Then the continuity equation

$$\nabla \cdot \boldsymbol{u} = 0 \tag{C.3}$$

yields

$$\nabla^2 \phi = 0 \tag{C.4}$$

For inviscid flow on the boundary the normal velocity, or normal derivative of ϕ, is zero. Equation (C.4) can be solved by separation of variables if the boundary can be described by a separable coordinate system.

The inviscid N–S equation (Euler equation) is

$$\boldsymbol{u}_t + (\boldsymbol{u} \cdot \nabla)\boldsymbol{u} = -\frac{1}{\rho}\nabla p + \boldsymbol{f} \tag{C.5}$$

© The Editor(s) (if applicable) and The Author(s), under exclusive license to Springer Nature Switzerland AG 2024
C. Y. Wang, *Essential Analytic Laminar Flow*, Synthesis Lectures on Engineering, Science, and Technology, https://doi.org/10.1007/978-3-031-36449-5

In the identity

$$(u \cdot \nabla)u = \frac{1}{2}\nabla|u|^2 - u \times (\nabla \times u) \tag{C.6}$$

the last term is zero due to Eq. (C.2). If the body force is conservative, such as gravity, $f = -\nabla V$. Equation (C.5) gives

$$\nabla\left(\frac{p}{\rho} + \frac{1}{2}|u|^2 + \phi_t + V\right) = 0 \tag{C.7}$$

We obtain Bernoulli's equation

$$\frac{p}{\rho} + \frac{1}{2}|\nabla\phi|^2 + \phi_t + V = C(t) \tag{C.8}$$

If the flow is steady and $f = 0$, then maximum pressure is p_0 at the stagnation point.

$$p = p_0 - \frac{\rho}{2}|u|^2 \tag{C.9}$$

C.1   Two-Dimensional Flow

Let (u, v) be velocities in the Cartesian (x, y) directions respectively. From Eq. (C.3) define a stream function ψ

$$u = \psi_y = \phi_x, \quad v = -\psi_x = \phi_y \tag{C.10}$$

These Cauchy-Riemann relations show the curves of ϕ and ψ are orthogonal. The steam function ψ also satisfies the Laplace equation and may be solved by separation of variables

$$\nabla^2\psi = 0 \tag{C.11}$$

For two-dimensional flow there exists a complex potential w

$$\phi + i\psi = w(z), \quad z = x + iy \tag{C.12}$$

Any analytic function of z, or the transform of z to another complex plane, satisfies the Laplace equation.

The complex potential for the flow in a corner of angle α is $w = Cz^{\pi/\alpha}$. Those for a source and a (irrotational) vortex are

$$w = \frac{m}{2\pi}\ln(z), \quad w = -\frac{i\Gamma}{2\pi}\ln(z) \tag{C.13}$$

where m is the source strength and $\Gamma = \oint \mathbf{u} \cdot d\mathbf{s}$ is the circulation.

The circle theorem states: If $w = f(z)$ is a flow with no solid boundaries and the singularities of f are at a distance greater than a in the complex z plane, then if a circle (cylinder) $|z| = a$ is introduced, the new complex potential is

$$w = f(z) + \overline{f}\left(\frac{a^2}{z}\right) \tag{C.14}$$

where the overbar denotes the complex conjugate.

Example C.1 The complex potential for uniform flow V in the y direction is $w = -iVz$. Thus the vertical potential flow over a circular cylinder is

$$w = -iVz + iV\left(\frac{a^2}{z}\right) \tag{C.15}$$

Equation (C.15) gives

$$\psi = -Vx + \frac{Va^2x}{(x^2+y^2)} = -Vr\cos(\theta)\left(1 - \frac{a^2}{r^2}\right) \tag{C.16}$$

where (r, θ) are cylindrical coordinates. Equation (C.16) can also be obtained by separation of variables in cylindrical coordinates. Figure C.1a shows the streamlines.

The transform

$$z = z' + \frac{b^2}{z'} \tag{C.17}$$

transforms the circle $|z'| = b$ into a slit and the circle $|z'| = a, \quad a > b$ into an ellipse in the z plane

$$\frac{x^2}{\left(a + \frac{b^2}{a}\right)^2} + \frac{y^2}{\left(a - \frac{b^2}{a}\right)^2} = 1 \tag{C.18}$$

The potential flow perpendicular to the major axis of an ellipse can be obtained by successive transforms. Starting with Eq. (C.15) in the z' plane

$$w = -iVz' + iV\left(\frac{a^2}{z'}\right) \tag{C.19}$$

Then substitute from Eq. (C.17)

$$z' = \frac{z}{2} + \sqrt{\frac{z^2}{4} - b^2} \tag{C.20}$$

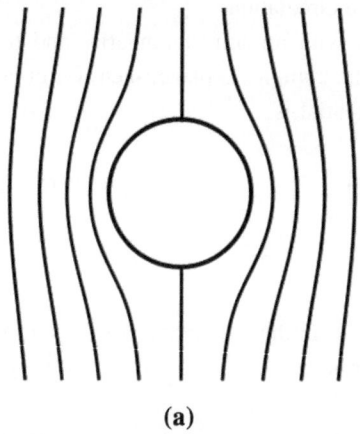 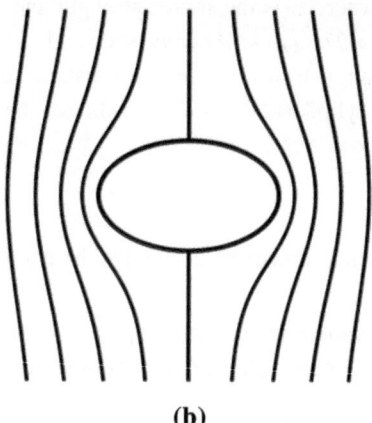

(a)   (b)

Fig. C.1 a Potential flow over a circular cylinder. **b** Flow over an elliptic cylinder (aspect ratio is 0.6)

Notice the appropriate sign is chosen to ensure the same uniform flow at infinity. The streamlines are shown in Fig. (C.1b). For potential flow, the flow direction can be reversed.

It can be shown that source strength and vortex strength are invariant under successive transforms.

C.2   Three-Dimensional Axisymmetric Flow

In three-dimensional Cartesian coordinates, the potential for a source is

$$\phi = -\frac{m}{\sqrt{x^2 + y^2 + z^2}} = -\frac{m}{\varrho} \tag{C.21}$$

where ϱ is the distance from the origin. The velocity m/ϱ^2 is radial and the total flux is $4\pi m$.

Let (u, w) be velocities in axisymmetric cylindrical coordinates (r, z). The potential equation is

$$\left(\frac{\partial^2}{\partial r^2} + \frac{1}{r} \frac{\partial}{\partial r} + \frac{\partial^2}{\partial z^2} \right) \phi = 0 \tag{C.22}$$

Equation (1.43) shows the stream function and potential function are related by

$$w = \phi_z = \frac{1}{r} \psi_r, \quad u = \phi_r = \frac{1}{r} \psi_z \tag{C.23}$$

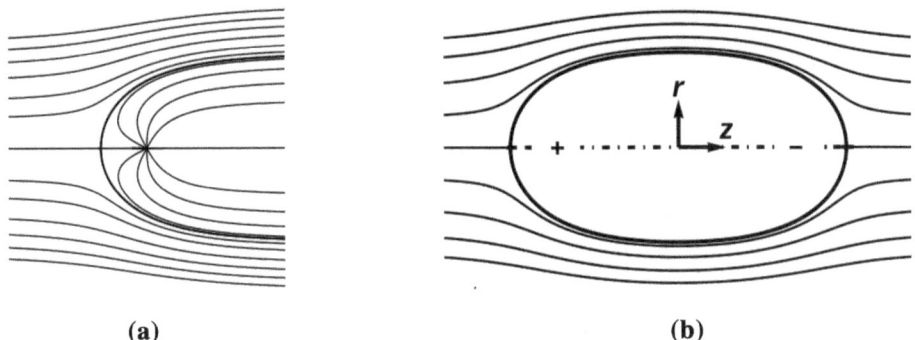

(a)   (b)

Fig. C.2 **a** Streamlines for uniform flow over a source. The thicker curve is the finger-like Rankine solid. **b** Streamlines for uniform flow over a finite Rankine solid

Thus the potential and the stream function of a source is

$$\phi = -\frac{m}{\sqrt{r^2 + z^2}}, \quad \psi = -\frac{mz}{\sqrt{r^2 + z^2}} \tag{C.24}$$

For uniform flow U in the z direction,

$$\phi = Uz, \quad \psi = \frac{U}{2}r^2 \tag{C.25}$$

The superposition of a uniform flow and a source give

$$\psi = \frac{U}{2}r^2 - \frac{mz}{\sqrt{r^2 + z^2}} = \frac{U}{2}r^2 - \frac{Ub^2 z}{\sqrt{r^2 + z^2}} \tag{C.26}$$

where b is the distance from the origin to the stagnation point on the semi-infinite Rankine solid shown in Fig. C.2a.

For a finite Rankine solid, put a sink of same strength at the rear of the source. The stream function is

$$\psi = \frac{U}{2}r^2 - \frac{m(z + a)}{\sqrt{r^2 + (z + a)^2}} + \frac{m(z - a)}{\sqrt{r^2 + (z - a)^2}} \tag{C.27}$$

Typical streamlines are shown in Fig. C.2(b).

As a approaches zero, Eq. (C.27) becomes uniform flow over a doublet

$$\psi = \frac{U}{2}r^2 - \frac{\sigma r^2}{(r^2 + z^2)^{\frac{3}{2}}} = \frac{U}{2}r^2 \left[1 - \frac{b^3}{(r^2 + z^2)^{\frac{3}{2}}} \right] \tag{C.28}$$

Equation (C.28) is also the stream function for the potential flow over a sphere of radius b.

By a distribution of sources and sinks the potential flow over any axisymmetric body can be obtained using Rankine's method.

Exercises

C.1 Obtain the stream function for a two-dimensional Rankine solid similar to Eq. (C.27)
C.2 Find the axisymmetric potential flow towards a cone.

Notes

Best sources for potential flow are Milne-Thomson (2011), Yih (1977). Also see Milne-Thomson's book for the potential flow over ellipsoids.

References

L.M. Milne-Thomson, *Theoretical Hydrodynamics*, 5th edn (Dover, NY, 2011)

C.S. Yih, *Fluid Mechanics* (West River Press, MI, 1977)

Index